ANNALS *of* THE NEW YORK ACADEMY OF SCIENCES

ISSUE

Blavatnik Awards for Young Scientists 2012

The New York Academy of Sciences Blavatnik Awards for Young Scientists acknowledges and celebrates the excellence of our most noteworthy young scientists and engineers in New York, New Jersey, and Connecticut. The awards recognize highly innovative, impactful, and interdisciplinary accomplishments in the life sciences, physical sciences, mathematics, and engineering with unrestricted financial prizes for both finalists and awardees.

TABLE OF CONTENTS

About the Blavatnik Family Foundation

The Blavatnik Family Foundation is an active supporter of many leading educational, scientific, cultural, and charitable institutions in the United States, the United Kingdom, Israel, and throughout the world. Recipients of foundation support include Oxford University, Harvard University, Tel Aviv University, the Royal Opera House, the Hermitage, the National Portrait Gallery, the British Museum, the National Gallery of Art, the Metropolitan Museum of Art, the New York Academy of Sciences, the White Nights Foundation of America, numerous Jewish charitable organizations, and countless other philanthropic institutions. The foundation is headed by Len Blavatnik, an American industrialist, who is the founder and chairman of Access Industries, a privately held industrial group with global interests in natural resources and chemicals, media and telecommunications, and real estate.

BLAVATNIK FAMILY FOUNDATION

The New York Academy of Sciences

Published by Blackwell Publishing
On behalf of the New York Academy of Sciences

Boston, Massachusetts
2013

Ann. N.Y. Acad. Sci. ISSN 0077-8923

ANNALS OF THE NEW YORK ACADEMY OF SCIENCES
Issue: *Blavatnik Awards for Young Scientists 2012*

What can stimulated emission do for bioimaging?

Lu Wei and Wei Min

Department of Chemistry, Columbia University, New York, New York

Address for correspondence: Wei Min, Department of Chemistry, Columbia University, 3000 Broadway, New York, NY 10027. wm2256@columbia.edu

Advances in bioimaging have revolutionized our ability to study life phenomena at a microscopic scale. In particular, the stimulated emission process, a universal mechanism that competes with spontaneous emission, has emerged as a powerful driving force for advancing light microscopy. The present review summarizes and compares three related techniques that each measure a different physical quantity involved in the stimulated emission process in order to tackle various challenges in light microscopy. Stimulated emission depletion microscopy, which detects the residual fluorescence after quenching, can break the diffraction-limited resolution barrier in fluorescence microscopy. Stimulated emission microscopy is capable of imaging nonfluorescent but absorbing chromophores by detecting the intensity gain of the stimulated emission beam. Very recently, stimulated emission reduced fluorescence microscopy has been proposed, in which the reduced fluorescence due to focal stimulation is measured to extend the fundamental imaging-depth limit of two-photon microscopy. Thus, through ingenious spectroscopy design in distinct microscopy contexts, stimulated emission has opened up several new territories for bioimaging, allowing examination of biological structures that are ever smaller, darker, and deeper.

Keywords: stimulated emission; superresolution; nonfluorescent chromophore; pump-probe microscopy; deep tissue imaging; imaging-depth limit

Introduction

Light microscopy, since its invention several centuries ago, has played an indispensable role in the life sciences to unveil valuable spatial and temporal information in the study of cells, tissues, and organisms.[1–3] Advances in light microscopy make visualization of live cell composition, dynamics, and physiology possible at a microscopic scale. Among the conceptual and technical factors that have propelled the development of modern light microscopy, stimulated emission is one of the current frontiers. The existence of the stimulated emission process was first theoretically postulated by Einstein back in 1917. It was later confirmed experimentally and now is understood as a universal optical process in which a molecule at its excited state can be stimulated down to its ground state by an incident photon with proper frequency, simultaneously creating a new coherent photon with the same phase, frequency, polarization, and direction as the incident

one. Figure 1 illustrates the competition between stimulated emission and spontaneous emission (i.e., fluorescence) processes.

The first and arguably the most notable application of stimulated emission in bioimaging is stimulated emission depletion (STED) microscopy for breaking the diffraction-limited spatial resolution of lens-based far-field fluorescence microscopy. Since its original proposal in 1994, STED has extended fluorescence microscopy to nanoscopy.[4,5] Going beyond the popular fluorescence contrast, stimulated emission was applied in 2009 to the detection and imaging of absorbing chromophores with nondetectable fluorescence using pump-probe optical techniques.[6] Most recently, inspired by STED microscopy and stimulated emission microscopy, stimulated emission reduced fluorescence (SERF) microscopy was proposed to extend the fundamental imaging-depth limit of two-photon fluorescence microscopy inside highly scattering samples.[7] Thus, when applied in different microscopy contexts,

doi: 10.1111/nyas.12079

Ann. N.Y. Acad. Sci. 1293 (2013) 1–7 © 2013 New York Academy of Sciences.

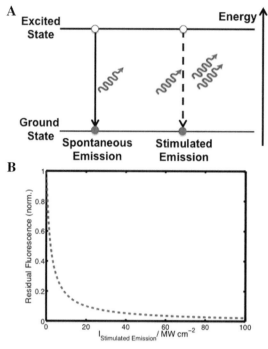

Figure 1. Comparison between the spontaneous emission and the stimulated emission. (A) For spontaneous emission, upon excitation, the molecule will relax from its excited state back to its ground state and concurrently emit a fluorescence photon at a certain frequency with a random phase and direction. In contrast, in stimulated emission, the molecule, after being excited to its excited state, also experiences the incoming photon(s) whose energy matches the energy gap between its excited and ground state. This molecule is then brought back to the ground and emits a photon exhibiting the identical physical properties as the incident stimulated emission photons. (B) Depletion of the fluorescence from a typical fluorophore as a function of the laser intensity I of a continuous-wave stimulated emission beam, illustrating the competition between the spontaneous emission and the stimulated emission.

stimulated emission has opened up several frontiers of bioimaging, allowing one to look at target structures that are much smaller, darker, and deeper than previously possible. We will review these three techniques individually before summarizing their underlying interconnections.

Breaking the diffraction-limited spatial resolution

In lens-based far-field fluorescence microscopy, diffraction-limited spatial resolution (∼200 nm) has been a serious issue for over a century.[8] Because light cannot be focused tighter than its diffraction limit, the image of an object that is smaller than

the diffraction limit inevitably becomes blurry. As a result, the ability to clearly resolve fine biological structures smaller than 200 nm was exclusive to electron microscopy, which is, however, not compatible with live imaging. Since its original theoretical description in 1994[4] and first experimental demonstration in 1999,[9] STED microscopy has achieved lateral resolution of about 15–20 nm in biological samples, which is a 10- to 12-fold increase over diffraction-limited resolution,[10] opening up the entire super-resolution field for resolving structures that are too blurry for conventional fluorescence microscopy.

STED microscopy typically adopts a confocal scheme, using a tightly focused excitation laser spot, collinearly combined with a doughnut-shaped and red-shifted STED laser beam (Fig. 2A) to scan across the sample plane. The rationale behind STED is the selective deactivation of fluorophores at the edge of a laser spot via stimulated emission depletion, thereby allowing only the fluorophores at the very center to fluoresce[11] (Fig. 2B). By elevating the intensity of the doughnut-shaped STED beam to saturate the stimulated emission process, resolution can be continuously reduced into a progressively fine scale, extending the classic Abbe's diffraction limit of $d = \lambda/(2NA)$ to a new diffraction-unlimited regime of $d \approx \lambda/(2NA\sqrt{1 + I/I_s})$, where d is the full-width-at-half-maximum (FWHM) of the focal spot at the focal plane, λ is the excitation wavelength, NA is the numerical aperture, I_s is the intensity at which half of the fluorophores are quenched (i.e., loss of fluorescence due to stimulated emission), and I is the applied intensity of the doughnut laser beam.[12]

In recent years, STED has matured into a popular and widely used superresolution technique, especially with the advent of continuous wave (CW) STED.[13] Structural analysis of the structures and distributions of proteins such as tubulin and other cytoskeletal filaments on suborganelle levels has become standard using STED microscopy.[14] Multicolor STED has also been made possible, from its first demonstration on colocalization imaging of synaptic and mitochondrial protein clusters with 5 nm precision[15] to more complicated analyses, such as the analysis of protein–protein interaction in parallel channels.[16,17] Moreover, with the improvement of genetically encoded fluorescent proteins,[18,19] live-cell STED microscopy has

A

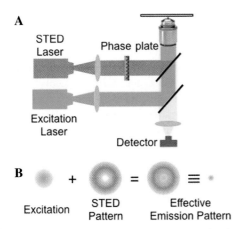

B

Figure 2. Setup and principal illustration of stimulated emission depletion (STED) microscopy. (A) Basic setup of STED microscopy. A spatially shaped doughnut STED beam created by using a phase plate is collinearly combined with an excitation beam. Both beams are tightly focused onto the sample. A detector detects the residual fluorescence signal in the presence of the STED laser beam. (B) An illustration of the pattern of each focused laser beam onto sample as well as the final effective excitation profile achieved by the STED design. The diffraction-limited excitation beam (green) is overlapped with the doughnut-shaped STED beam (red) that quenches the fluorescence at the edge of the excitaton beam. The remaining fluorescence is only generated at the center of the excitation beam, which effectively results in an overall narrower emission pattern.

offered rich and valuable information. For example, the time-lapse STED imaging of both dendritic spine cells[20] and living mouse brains[21] exhibit important structural details. Additionally, STED imaging of cultured hippocampal neurons illustrates that endosomal sorting of synaptic vesicles is a rapid pathway.[22] In addition, examination of membrane lipid dynamics at nanoscale reveals that unlike phosphoglycerolipids, sphingolipids and glycosylphosphatidylinositol-anchored proteins are transiently ($\sim$10–20 ms) trapped in cholesterol-mediated molecular complexes dwelling within areas of less than 20 nm diameter.[23] Efforts have also been devoted to enable video-rate STED ($\sim$28 frames/s),[24] which is important for imaging dynamic processes such as synaptic plasticity. Furthermore, in order for deep tissue superresolution imaging, aberration-reducing optics have been utilized to demonstrate a resolution of 60–80 nm in living organotypic brain slices at depths of up to 120 μm.[25] Meanwhile, two-photon excitation in combination with CW STED beam has also been

demonstrated to be feasible[26] and has been applied to image brains slices with a threefold resolution increase at below 100 μm depth.[27]

STED microscopy is undoubtedly a milestone in the development of advanced fluorescence microscopy. Its robust and general spectroscopic mechanism and intrinsic compatibility with scanning confocal and multiphoton microscopy make it widely useful in various fields of biomedical sciences. However, because of the high stimulated emission laser intensity, a certain degree of photodamage on the samples is inevitable. In addition, one certainly has to acquire some optics expertise before building a complex STED microscope, but once built, it can be used as easily as a confocal microscope.[11]

Imaging nonfluorescent but absorbing chromophores

In addition to its ability to increase spatial resolution of fluorescence microscopy for a sharper visualization of what is being seen, stimulated emission has also been applied to detection and imaging of nonfluorescent but absorbing chromophores to explore what appears to be invisible.[6] There exist many chromophores in life systems that absorb heavily but have undetectable fluorescence, such as hemoglobin and cytochromes, because of their short excited state lifetimes ($<$1 ps) due to their rapid nonradiative decay processes over spontaneous emission.[28] Unfortunately, direct one-laser absorption microscopy has low sensitivity. Hence, imaging these nonfluorescent but absorbing molecules with sufficient sensitivity has been rather challenging, especially in complex biological samples.

Stimulated emission microscopy adopting a high-frequency modulation transfer scheme provides a suitable solution for the above challenge. Figure 3 shows a cartoon of the stimulated emission microscopy setup and its signal generation process. A pulsed (pulse width $\sim$200 fs) excitation laser beam is spatially overlapped and temporally synchronized with another pulsed ($\sim$200 fs) stimulated emission beam whose wavelength is properly red shifted. A few hundred femtoseconds of time delay between these two pulse trains is chosen to prevent the occurrence of instantaneous optical processes, such as, for example, stimulated Raman scattering[29,30] and two-photon absorption.[31] In the common focal volume, after being excited

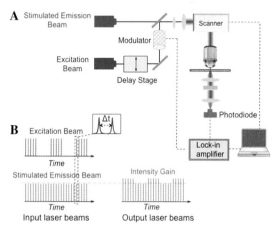

Figure 3. Cartoon demonstration of a stimulated emission microscopy setup for imaging nonfluorescent chromophores. (A) A modulated excitation laser beam is collinearly aligned with a stimulated emission beam before being sent into the sample. The intensity gain of the transmitted stimulated emission beam is then demodulated by a lock-in amplifier at the modulation frequency to ensure shot-noise limited detection sensitivity. (B) Temporal characterization of the input and output beams. The modulated excitation pulse train (blue) is synchroized with the stimulated emission pulse train (red) with a temporal time delay (∼300 fs) between these two pulse trains. With the excitation beam on, the stimulated emission pulse train experiences an intensity gain; with the excitation beam off, the stimulated emission pulse train intensity remains unchanged after interacting with the sample.

by the excitation pulses, the chromophores are subsequently interrogated by the stimulated emission pulses, which bring them down to the ground state faster than the nonradiative process. As a result, after passing through samples, the intensity of the stimulated emission beam will be increased due to photons newly created by the chromophores. However, the relative intensity gain is normally small ($<10^{-3}$) that can easily be buried by the low-frequency laser noise. To achieve the necessary detection sensitivity, a high-frequency modulation transfer scheme is employed, where the excitation beam intensity is modulated by an acousto- or electro-optic modulator at high frequency (>1 MHz) and the transmitted stimulated emission beam is demodulated by a lock-in amplifier at the same modulation frequency to extract the intensity gain and to reject noises at other frequencies.

Stimulated emission microscopy has been successfully applied to visualizing chromoproteins (the nonfluorescent variants of the green fluorescent protein; GFP) in live *Escherichia coli* cells, mon-

itoring lacZ gene expression with a chromogenic reporter, and mapping transdermal drug distributions without histological sectioning.[6] The detection limit for stimulated emission microscopy is 60 nM for crystal violet with one-second integration time. This sensitivity effectively corresponds to a few (∼5) molecules in focus.[6] The advantages of stimulated emission microscopy include: (1) because of its signal dependence on both excitation and stimulated emission laser intensities, its nonlinear nature offers intrinsic 3D sectioning ability; (2) the high-frequency modulation scheme ensures shot-noise limited detection sensitivity by getting rid of the lower frequency laser noise; and (3) the modulation transfer between two laser beams (in comparison with traditional one-beam absorption microscopy) avoids undesired signal artifacts from heterogeneous sample scattering.[32] These features make stimulated emission microscopy a desirable technique for imaging chromophores with high sensitivity and specificity in complex biological environments. The major complication for stimulated emission microscopy is the difficulty of synchronizing the two-femtosecond laser pulse trains.

Extending the fundamental imaging-depth limit

It is generally believed that the spatial resolution and the penetration depth of a given imaging modality are inversely correlated. For example, MRI has a poor spatial resolution but superb penetration depth. In the domain of light microscopy, while the diffraction-limited spatial resolution barrier has been broken by STED, photoactivated localization microscopy (PALM), and stochastic optical reconstruction microscopy (STORM),[4,5,33,34] the deepest penetration into scattering samples with subcellular resolution is achieved currently by two-photon fluorescence microscopy.[35] By employing a nonlinear optical excitation, two-photon fluorescence is primarily generated at the focal volume. Such a spatially confined excitation scheme thus permits the capture of fluorescence photons emitted and then scattered from the focus by a nondescanned detector, dramatically increasing the detection sensitivity. This profound advantage of two-photon imaging leads to an imaging depth that is more than three times deeper than what can be achieved with confocal microscopy.

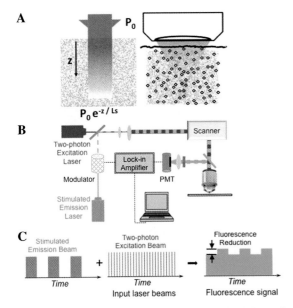

Figure 4. Extending the fundamental imaging-depth limit of two-photon microscopy by stimulated emission reduced fluorescence (SERF) microscopy. (A) Illustration of the fundamental imaging depth limit of two-photon microscopy in the scattering tissue sample. Incident light power decreases exponentially into the scattering sample. Thus, the fundamental imaging depth limit of two-photon microscopy is defined as the depth at which the in-focus signal and the out-of-focus background signal are equal. (B) The proposed SERF microscope setup. A modulated stimulated emission beam is combined collinearly with a two-photon excitation beam. Then, the reduced fluorescence due to quenching is measured by a lock-in amplifier and used to form SERF images. (C) The modulation transfer scheme of SERF. The CW stimulated emission beam is modulated at a high frequency (> MHz). When combined with the two-photon excitation pulse train, this modulated stimulated emission beam leads to the resulting two-photon fluorescence signal modulated at the same frequency.

However, a fundamental imaging-depth limit still exists for two-photon microscopy. For example, for mouse brain tissues labeled with GFP, the corresponding depth limit is about 1 mm.[36] Such a depth limit is not constrained by the available laser power, but instead by the achievable image contrast.[36–38] When imaging highly scattering samples, with the increase of focusing depth, the laser power has to be elevated accordingly in order to compensate for scattering loss. At some point (Fig. 4A), the laser power deposited at sample surfaces becomes so strong that it generates comparable or even stronger fluorescence than that from the focal volume, which thus deteriorates the achievable image contrast. Formally, the depth where the in-focus

signal and the out-of-focus background are equal to each other is defined as the fundamental imaging-depth limit.[36,38] Thus, the conventional optical sectioning picture that two-photon fluorescence is generating only within the focal volume breaks down here. Obviously, further increasing laser power cannot extend this imaging-depth limit.

One plausible strategy to extend the fundamental imaging-depth limit is to suppress the out-of-focus background as recently demonstrated by focal switching of optical highlighter fluorescent probes.[39–41] This strategy requires the use of special fluorescent probes, which can sometimes be inconvenient. To be generally applicable to all fluorophores, stimulated emission reduced fluorescence (SERF) microscopy is proposed to introduce an additional stimulated emission process to distinguish the desired focal signal from the out-of-focus background:[7] when the stimulated emission laser is focused collinearly with a two-photon laser, the stimulated emission process will preferably occur at the focus because the intensity of the stimulated emission beam at the focus is much higher than its out-of-focus counterpart. Figure 4B depicts the proposed SERF microscope setup. The technical aspect of SERF is analogous to optical lock-in detection (OLID)[42] and synchronously amplified fluorescence image recovery (SAFIRe)[43] techniques; however, OLID and SAFIRe tackle problems related to autofluorescence background.

By spatially overlapping the two-photon excitation laser (pulse train) with an intensity-modulated and red-shifted CW-stimulated emission laser (pulsed laser works equally well, but is technically more demanding), the two-photon excited fluorescence is collected and then demodulated with a lock-in amplifier at the modulation frequency. Instead of detecting the residual fluorescence, as in STED, or the transmitted stimulated emission beam, as in stimulated emission microscopy, SERF effectively detects the fluorescence signal reduction (Fig. 4C) due to stimulated emission–based quenching. Thus, SERF combines the fluorescence quenching mechanism, as in STED, and the high-frequency modulation transfer scheme, as in stimulated emission microscopy. It can be quantitatively demonstrated that, at the weak fluorescence quenching region (i.e., where the fluorescence signal reduction has a linear relationship to the applied stimulated emission laser intensity), the final SERF signal exhibits an overall three-photon dependence,[7] including the original two-photon excitation and the new one-photon stimulated emission. As shown by the numerical simulation,[7] such a higher order nonlinearity (compared to the standard second-order effect) offers a much higher signal-to-background ratio and a contrast-limited imaging-depth limit that is effectively 1.8 times deeper. The disadvantage of SERF lies mainly in its imaging speed: because of the relatively moderate signal size deep inside scattering samples, a relatively long integration time may be needed for acquiring a decent signal-to-noise ratio.

Conclusion

We now summarize the interconnection of the above three stimulated emission–based bioimaging techniques. There exist three total measurable physical quantities in the stimulated emission process: (1) the residual fluorescence; (2) the reduced fluorescence; and (3) the enhanced stimulated emission beam. As illustrated in Figure 5, each of the techniques discussed above measures one of these three quantities to accomplish the respective goal. To squeeze the effective emission pattern, STED measures the residual fluorescence in the center of the focal spot quenched by the doughnut-shaped stimulated emission depletion beam. To generate an optically detectable signal from nonfluorescent chromophores, stimulated emission microscopy measures the enhanced intensity of the transmitted stimulated emission beam. To create even higher order nonlinearity on top of two-photon excited fluorescence, SERF measures the reduced fluorescence by introducing an additional stimulated emission laser beam.

Since both STED and SERF deal with fluorescent molecules, it is highly constructive to compare the technical aspects of these two. Both techniques harness the fluorescence quenching process of stimulated emission. However, STED aims to break the spatial resolution limit while SERF is designed to extend the penetration depth. It is interesting to see how stimulated emission can contribute to both ends. Third, the stimulated emission beam in STED is spatially shaped, whereas it is being temporally modulated in SERF. Finally, STED works best in the fluorescence depletion region where the stimulated emission intensity is strong, while SERF operates in the nonsaturating region where the stimulated

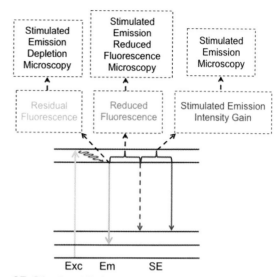

SE: Stimulated Emission; Exc: Excitation; Em: Emission

Figure 5. Diagram of physical quantities in the stimulated emission process that are being used in the three imaging techniques reviewed here. In the presence of stimulated emission, the three measurable optical quantities are the residual fluorescence; the reduced fluorescence; and the enhanced stimulated emission beam. Stimulated emission depletion microscopy (STED) makes use of the residual fluorescence to improve the spatial resolution problem; stimulated emission reduced fluorescence microscopy detects the reduced fluorescence to extend the fundamental imaging-depth limit of two-photon microscopy; and stimulated emission microscopy detects the intensity gain of the transmitted stimulated emission laser beam to image nonfluorescent chromophores.

emission intensity is moderate in order to perform deep imaging.

There is no doubt that the universal stimulated emission principle has played a significant role in driving the development of advanced bioimaging techniques, allowing one to look at target structures that are much smaller, darker, and deeper than previously possible. It is rather striking to see how it can be applied in distinct microscopy contexts to bring novel solutions to seemingly unrelated problems. More exciting biomedical applications in various areas are expected for many years to come.

Acknowledgments

The authors thank Zhixing Chen and Rafael Yuste for helpful discussions. W.M. acknowledges the start-up funds from Columbia University, and grant support from the Kavli Institute for Brain Science.

Conflicts of interest

The authors declare no conflicts of interest.

References

1. Pawley, J.B. (ed.) 2006. *Handbook of Biological Confocal Microscopy 3rd edn.* New York: Springer.
2. Lakowicz, J.R. 1983. *Principles of Fluorescence Spectroscopy.* New York: Plenum Press.
3. Rosenthal, C.K. *et al.* 2009. Nature milestones in light microscopy. *Nat. Cell Biol.* **11:** 1165.
4. Hell, S.W. & J. Wichmann. 1994. Breaking the diffraction resolution limit by stimulated emission: stimulated-emission-depletion fluorescence microscopy. *Opt. Lett.* **19:** 780–782.
5. Hell, S.W. 2007. Far-field optical nanoscopy. *Science* **316:** 1153–1158.
6. Min, W. *et al.* 2009. Imaging chromophores with undetectable fluorescence by stimulated emission microscopy. *Nature* **461:** 1105–1109.
7. Wei, L., Z. Chen & W. Min. 2012. Stimulated emission reduced fluorescence microscopy: a concept for extending the fundamental depth limit of two-photon fluorescence imaging. *Biomed. Opt. Express* **3:** 1465–1475.
8. Abbe, E. 1873. Beiträge zur Theorie des Mikroskops und der mikroskopischen Wahrnehmung. *Arch. Mikr. Anat.* **9:** 413–468.
9. Klar, T.A. & S.W. Hell. 1999. Subdiffraction resolution in far-field fluorescence microscopy. *Opt. Lett.* **24:** 954–956.
10. Donnert, G. *et al.* 2006. Macromolecular-scale resolution in biological fluorescence microscopy. *Proc. Natl. Acad. Sci. USA* **103:** 11440–11445.
11. Hell, S.W. 2009. Microscopy and its focal switch. *Nat. Methods* **6:** 24–32.
12. Hell, S.W., S. Jakobs & L. Kastrup. 2003. Imaging and writing at the nanoscale with focused visible light through saturable optical transitions. *Appl. Phys. A.* **77:** 859–860.
13. Willig, K.I. *et al.* 2007. STED microscopy with continuous wave beams. *Nat. Methods* **4:** 915–918.
14. Kasper, R. *et al.* 2010. Single-molecule STED microscopy with photostable organic fluorophores. *Small* **6:** 1379–1384.
15. Donnert, G. *et al.* 2007. Two-color far-field fluorescence nanoscopy. *Biophys J.* **92:** L67–L69.
16. Pellett, P.A. *et al.* 2011. Two-color STED microscopy in living cells. *Biomed. Opt. Express* **2:** 2364–2371.
17. Bückers, J. *et al.* 2011. Simultaneous multi-lifetime multi-color STED imaging for colocalization analyses. *Opt. Express* **19:** 3130–3143.
18. Chalfie, M. *et al.* 1994. Green fluorescent protein as a marker gene expression. *Science* **263:** 802–805.
19. Heim, R., A.B. Cubitt & R.Y. Tsien. 1995. Improved green fluorescence. *Nature* **373:** 663–664.
20. Nägerl U.V. *et al.* 2008. Live-cell imaging of dendritic spines by STED microscopy. *Proc. Natl. Acad. Sci. USA* **105:** 18982–18987.
21. Berning, S. *et al.* 2012. Nanoscopy in a living mouse brain. *Science* **335:** 551.
22. Hoopmann, P. *et al.* 2010. Endosomal sorting of readily releasable synaptic vesicles. *Proc. Natl. Acad. Sci. USA* **107:** 19055–19060.
23. Eggeling, C. *et al.* 2009. Direct observation of the nanoscale dynamics of membrane lipids in a living cell. *Nature* **457:** 1159–1162.
24. Westphal, V. *et al.* 2008. Video-rate far-field optical nanoscopy dissects synaptic vesicle movement. *Science* **320:** 246–249.
25. Urban, N.T. *et al.* 2011. STED nanoscopy of actin dynamics in synapses deep inside living brain slices. *Biophys. J.* **101:** 1277–1284.
26. Moneron, G. & S.W. Hell. 2009. Two-photon excitation STED microscopy. *Opt. Express* **17:** 14567–14573.
27. Ding, J.B., K.T. Takasaki & B.L. Sabatini. 2009. Supraresolution imaging in brain slices using stimulated-emission depletion two-photon laser scanning microscopy. *Neuron* **63:** 429–437.
28. Turro, N.J. 1991. *Modern Molecular Photochemistry.* California: University Science Books.
29. Freudiger, C.W. *et al.* 2008. A label-free biomedical imaging with high sensitivity by stimulated Raman scattering microscopy. *Science* **322:** 1857–1861.
30. Saar, B.G. *et al.* 2010. Video-rate molecular imaging in vivo with stimulated Raman scattering. *Science* **330:** 1368–1370.
31. Fu, D. *et al.* 2007. High-resolution in vivo imaging of blood vessels without labeling. *Opt. Lett.* **32:** 2641–2643.
32. Wei, L. & W. Min. 2012. Pump-probe optical microscopy for imaging nonfluorescent chromophores. *Anal. Bioanal. Chem.* **403:** 2197–2202.
33. Huang, B., H. Babcock & X. Zhuang. 2010. Breaking the diffraction barrier: super-resolution imaging of cells. *Cell* **143:** 1047–1058.
34. Betzig, E. *et al.* Imaging intracellular fluorescent proteins at nanometer resolution. *Science* **313:** 1642–1645.
35. Denk, W., J.H. Strickler & W.W. Webb. 1990. Two-photon laser scanning fluorescence microscopy. *Science* **248:** 73–76.
36. Theer, P., M.T. Hasan & W. Denk. 2003. Two-photon imaging to a depth of $1000\,\mu$m in living brains by use of a Ti:Al$_2$O$_3$ regenerative amplifier. *Opt. Lett.* **28:** 1022–1024.
37. Helmchen, F. & W. Denk. 2005. Deep tissue two-photon microscopy. *Nat. Methods* **2:** 932–940.
38. Theer, P. & W. Denk. 2006. On the fundamental imaging-depth limit in two-photon microscopy. *J. Opt. Soc. Am. A. Opt. Image. Sci. Vis.* **23:** 3139–3149.
39. Kao, Y.-T. *et al.* 2012. Focal switching of photochromic fluorescent proteins enables multiphoton microscopy with superior image contrast. *Biomed. Opt. Express* **3:** 1955–1963.
40. Chen Z. *et al.* 2012. Extending the fundamental imaging-depth limit of multi-photon microscopy by imaging with photo-activatable fluorophores. *Opt. Express* **20:** 18525–18536.
41. Zhu, X., Y.-T. Kao & W. Min. 2012. Molecular-switch-mediated multiphoton fluorescence microscopy with high-order nonlinearity, *J. Phys. Chem. Lett.* **3:** 2082–2086.
42. Marriott, G. *et al.* 2008. Optical lock-in detection imaging microscopy for contrast-enhanced imaging in living cells. *Proc. Natl. Acad. Sci. USA* **105:** 17789–17794.
43. Richards, C.I., J.C. Hsiang & R.M. Dickson. 2010. Synchronously amplified fluorescence image recovery (SAFIRe). *J. Phys. Chem. B.* **114:** 660–665.

Ann. N.Y. Acad. Sci. ISSN 0077-8923

ANNALS OF THE NEW YORK ACADEMY OF SCIENCES

Issue: *Blavatnik Awards for Young Scientists 2012*

Plant invasions across the Northern Hemisphere: a deep-time perspective

Jason D. Fridley

Department of Biology, Syracuse University, Syracuse, New York

Address for correspondence: Jason D. Fridley, Syracuse University, Department of Biology, 107 College Place, Syracuse, NY 13244. fridley@syr.edu

Few invasion biologists consider the long-term evolutionary context of an invading organism and its invaded ecosystem. Here, I consider patterns of plant invasions across Eastern North America, Europe, and East/Far East Asia, and explore whether biases in exchanges of plants from each region reflect major selection pressures present within each region since the late Miocene, during which temperate Northern Hemisphere floras diverged taxonomically and ecologically. Although there are many exceptions, the European flora appears enriched in species well adapted to frequent, intense disturbances such as cultivation and grazing; the North American composite (Asteraceae) flora appears particularly well adapted to nutrient-rich meadows and forest openings; and the East Asian flora is enriched in shade-tolerant trees, shrubs, and vines of high forest-invasive potential. I argue that such directionality in invasions across different habitat types supports the notion that some species are preadapted to become invasive as a result of differences in historical selection pressures between regions.

Keywords: preadaptation; Eastern North America; naturalized plants; invasion biology

Introduction

Modern biologists are accustomed to treating the introduction and escape of an organism into a new ecosystem as an invasion, particularly in the wake of Charles Elton's classic work that helped define the vibrant subdiscipline of invasion biology.[1,2] As Elton was well aware, however, paleontologists have long considered the process of a species inhabiting a novel environment as a natural component of biotic exchange between regions of disparate evolutionary histories, with species migrating to new ecosystems as the result of repeated shifts of land masses and sea levels over the past 400 million years.[3,4] Although most invasions today are the result of human introduction rather than natural migration, they nonetheless occur within a global evolutionary context that has often produced organisms of strikingly different form and function for a given environmental setting.[5] Thus, modern invasions can be seen as only the latest example of a long history of biotic interchanges between global faunal and floristic regions.[6]

Against this backdrop, it is surprising that so few invasion biologists consider the deep time evolutionary histories of an introduced species and its incipient ecosystem, despite explicit calls for a research agenda that views invasions from the perspective of biotic exchanges between regions.[6,7] For example, studies designed to understand how a species becomes invasive in a given habitat typically focus on traits of the invader in relation to properties of the invaded community (e.g., Is there an empty niche or unconsumed resource? Is the invader lacking enemies that limit native populations?), without considering properties of the invader's native region that might preadapt it for success in a foreign environment. In the absence of a larger context that considers properties of both the invaded region and the source region of invaders, researchers may be ill-equipped to answer some of the most compelling questions of invasion biology: Why are some regions (e.g., tropics) so rarely invaded? Why are species native to certain regions (e.g., Europe) so well represented as global invaders? Why are certain lineages (e.g., *Pinus* in the Southern Hemisphere)

doi: 10.1111/nyas.12107

Ann. N.Y. Acad. Sci. 1293 (2013) 8–17 © 2013 New York Academy of Sciences.

particularly invasive? Why are certain species (e.g., *Acacia mearnsii, Lantana camara, Spartina anglica*) invasive no matter where they have been introduced?

Here, I explore the modern exchange of plants across the Northern Hemisphere (NH), focusing on linkages of other NH vascular floras to that of Eastern North America (ENA). My first objective is to identify, for certain broadly defined habitats, whether plant invasions across the NH are directional—that is, accounting for differences in introduction effort, are introduced species from one region more likely to become invasive in a given habitat type? A second objective is to ask whether such biases, when found, are consistent with differences in the evolutionary (deep time) histories of the invading organism and its invaded region. For the first objective, I build on previous analysis[8] that identified biases in invader provenance across different habitat types of the Eastern United States (EUS), updated here to include provenance data at the country or subcountry scale for over 2000 EUS-naturalized and invasive species, plus several recently published accounts (see below) of the naturalized and invasive floras of regions across Europe, East Asia, and the Russian Far East. Because information on invaders to ENA from Western North America is less reliable than intercontinental invasions and they are likely a relatively small component of the invasive flora, I restrict my analysis to ENA and Eurasia. For the secondary objective, I consider in broad terms the paleo-floristic records of ENA, Europe, and East Asia and particularly the history of closed (forested) versus open habitats (meadows, fields, woodlands, savannas) over the last 10–12 million years, the period during which strong floristic provincialism developed across the NH. In tying together modern invasion patterns and the evolutionary history of these floras, I then make the case that some species appear to be preadapted as invaders: that is, they have evolved superior adaptations to particular selection pressures (e.g., grazing) that ultimately drive their success in a new region.

I begin by briefly describing the history of modern plant introductions to ENA, to set the stage for an analysis of the invasiveness of species from particular floristic regions that accounts for biases in the number of plant introductions among regions.

Introduction of foreign plants to Eastern North America

Post-Columbian plant introductions into Eastern North America came in two waves, the first consisting of species almost entirely from Europe (between around 1500 and 1850 A.D.) and the second dominated by introductions from East Asia starting in the late 1800s.[9–11] Those of the first wave were concomitant with European settlement, including crops, forage species, and weeds associated with forest clearing, annual cultivation, grazing, and human habitation.[10] The earliest account of naturalized plants in New England, by the Englishman John Josselyn in 1671, emphasized their association with livestock,[9] and there would have been few native species in ENA suitable for productive forage, in part because most native grasses and small forbs are not evergreen or active in early spring or late autumn.[12] Woody species were a small minority of the introductions, and appeared first as fruit trees and a few ornamentals (e.g., *Buxus sempervirens*), followed by a more avid horticultural exchange between American and European gardens from 1750 to 1850.[13]

A second wave of introductions began in 1861 with the first direct shipment of Japanese plants to Eastern U.S. gardens.[13] In contrast to plants introduced in the centuries before, those of the late 19th and early 20th centuries were typically woody and sourced directly from collecting trips in China, Korea, and Japan.[13] Unlike the majority of the European introductions, many East Asian species became garden escapes within a few decades of their introduction.[9] Merhoff[9] and Boufford[10] describe two presentations given by the Harvard botanist M. L. Fernald on nonnative species in New England, one in 1905 and the other in 1940. The first focused almost exclusively on European species with little concern over invaders, while the second focused on escaped garden plants and growing concerns about the effects of invaders on natural areas. Today, the ENA naturalized flora can be largely described as two distinct floras: one European, present in ENA for several centuries and nearly exclusive to the transformed landscapes of human settlement; and another East Asian, relatively recent, and more likely to invade the region's forested natural areas.[8,10]

Eastern U.S. naturalizations by donor country

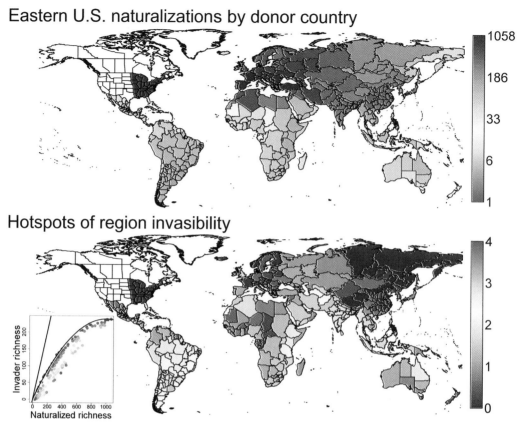

Figure 1. Top: the number of naturalized plants in the Eastern United States (green region) by country or region of origin, according to Ref. 15. Bottom: the number of invasive plants in the Eastern United States by country or region of origin, expressed as a function of the total number of naturalized plants from the same region (graph insert, lower left). A maximum constraint line (95th quantile) describing the relationship between country naturalized and invasive richness was determined as Invader richness = 0.447 × (Nat'zd richness) − 0.00237 × (Natz'd richness)² (curved line, insert; straight line is 1:1 relationship). Residuals from this line are expressed as standard deviations. Regions in red have the maximum number of invaders from their naturalized pool, while species of those regions in blue are less invasive than expected.

Global hotspots of where ENA invaders come from

Five hundred years of plant introductions, habitat modification, and landscape disturbance have led to a vast reorganization of plant communities across ENA. Although the major waves of introduction have come broadly from Europe and East Asia, it is an open question as to how the overall ENA flora is changing in the context of the global distribution of plants and ecosystems. Do the new dominant species come from particular regions, thus increasing the ecological similarity of ENA to specific foreign ecosystems?

To illustrate which regions of the globe are home to species that become particularly invasive after in-troduction into ENA, I present a geographic analysis of the compiled home ranges of all vascular plant species known to be naturalized in the Eastern U.S. region of ENA, as reported by USDA PLANTS,[14] using a database of native range occurrences compiled by the Germplasm Resources Information Network.[15] The naturalized flora is that reported by Fridley,[8] covering the extent of the eastern deciduous forest of North America[16] and the North American Atlantic floristic region,[17] including the states from Minnesota to Louisiana and eastward, excluding Florida (green region, Fig. 1). For each of the 2682 naturalized species reported by Fridley.[8] I assigned a list of home range occurrences, following the global geographic units of Hollis and Brummitt,[18] as depicted in Figure 1 (mostly at the country level, with larger countries divided into major provinces). I did not consider taxa at the subspecific or varietal level, and eliminated those species of unknown provenance or those of cultivation origin, resulting in a final dataset of the native geographic occurrences of 2238 naturalized species.

The top panel of Figure 1 shows the total number of EUS-naturalized plants by geographic region of origin, from a maximum of 1058 species native to Italy to zero species native to most oceanic islands. The current EUS-naturalized flora is dominated by European natives, particularly those from the northern section of the Mediterranean Basin from Spain to Turkey (Fig. 1). Few EUS-naturalized species come from the tropics or the Southern Hemisphere. Indeed, native provenances of the EUS-naturalized flora roughly follow from the history of plant introductions, being essentially Eurasian and predominantly European.[19]

A different story emerges, however, if the EUS-naturalized pool is restricted to those that have been reported as invasive by EUS management agencies, a subset of 449 species summarized by Fridley.[8] With all else being equal, regions that have contributed more naturalized species should also contribute more invaders, so a more interesting statistic than the number of EUS invaders native to each region is how much a region deviates from its expected number of invaders, given the size of its naturalized pool. It is not clear a priori what this relationship should be; a graph of the relationship (inset, Fig. 1) suggests the number of invaders from a region levels off after a linear increase with number of naturalized species. I fit a constraint line to these data as a quadratic 95th-quantile regression, and calculated residuals from this line as standard deviations. Regions in red show the empirical maximum number of invaders based on their overall naturalized pool, while species of those regions in blue are less invasive than expected. When shown this way, Europe and East/Far East Asia strongly stand out as regions whose naturalized plants are highly invasive in ENA, and plants of several other regions show a very low invasive tendency, including central Asia, the Indian subcontinent, and many Mediterranean countries. This approach to measuring region invasiveness has shortcomings, however, particularly for those countries that are the source of few naturalized species (central Africa, New Zealand) that appear as invader source hotspots

simply because invader richness is constrained by zero.

Interestingly, historical differences in the type of species introduced from Europe and East Asia—largely herbaceous in the former and woody in the latter—are also reflected in the invasiveness of species from these regions. This was addressed by Fridley,[8] who showed strong biases in the East Asian invader pool toward those invasive in forests and of woody growth form (39% of all East Asian invaders, compared to 23% and 24% of the EUS-native and EUS-naturalized floras, respectively). In the present analysis, woody species account for only 9% of the EUS invaders from many European countries, including Italy, France, Germany, Denmark, and the Ukraine. Further, the predominance of woody growth forms in the invasive species pools of East Asian regions extends to trees, shrubs, and vines: the highest percentages by region for trees are the Russian Far East (20–30%, e.g., Kuril Islands, Amur, Primorye) and Southeast Asia (Laos, Vietnam, Taiwan, Cambodia); those for shrubs include China (especially north central), Japan, and Korea (around 25%); and those for vines include Japan (12%) in addition to regions of far fewer naturalized species (e.g., South American subtropics). Further, of the 74 species invasive in the EUS that are endemic to East Asia, 64 are woody. In contrast, European invaders are closely associated with more open, regularly disturbed ecosystems and are overwhelmingly herbaceous (e.g., 79% of those from Italy).

ENA natives invading other regions: the case of meadow asters

Few habitats in ENA appear resistant to plant invasions, and broad classes of habitats—forests versus disturbed areas, for example—appear to be invaded by species from different geographic and evolutionary contexts. Is this true of the nonnative floras of other regions in the Northern Hemisphere? If many ENA plants have difficulty competing with European or East Asian species in their native habitats, are ENA species therefore less likely to be invasive in Europe or East Asia?

Across East Asia and the Russian Far East, where most nonnative floras have only recently been catalogued, ENA natives do appear to have made significant contributions to the nonnative pool, but only in particular habitats and with a strong taxonomic bias. For example, Weber et al.[20] list 33 species of invasive plants in China endemic to North America, nearly half (16) of which are composites (Asteraceae), substantially greater than the overall percentage of Asteraceae in the Chinese invasive flora (19%, or 52 species out of 270). This percentage is nearly equal to that of North American species in the Korean naturalized flora,[21] 32 (of 76) of which are in the Asteraceae, whereas in the whole naturalized flora Asteraceae accounts for 68 species out of 321 (21%). For the Russian Far East, Asteraceae composes the largest group of aliens (123 species, 18%), many of which are also from ENA.[22] Overall, Asteraceae species that are naturalized or invasive across much of East/Far East Asia appear to be common ENA meadow species, including species of *Solidago, Erigeron, Helianthus, Symphyotrichum* (formerly *Aster*), *Bidens, Coreopsis, Ambrosia, Conyza, Xanthium, Rudbeckia,* and *Ageratina* (formerly *Eupatorium*).

In contrast, woody species are poorly represented in East Asian nonnative floras, and the few that are represented are generally not native to ENA forests. Woody species compose less than 10% of the 270 invasive nonnative plants in China.[20] Only four woody species (*Acacia farnesiana, Rhus typhina, Parthenocissus quinquefolia, Robinia pseudo-acacia*) were reported native to ENA, and only the latter two invade forests or forest margins. Similarly, all but 10 of the 321 naturalized plants in Korea listed in a working list compiled by H.H.M. Lee (based on Refs. 21–25) are herbaceous, and only two of the woody species are from North America (*R. pseudo-acacia* and *Amorpha fruticosa*), the majority from elsewhere in Asia. Miyawaki and Washitani[26] noted a large contribution of North American species in Japanese riparian areas (37% of the 87 species reported as invasive aliens), but all except *R. pseudo-acacia* were herbaceous. Auld *et al.*[27] noted the overall lack of woody or forest invaders in Japan, including those present that have not reached the same degree of invasiveness as seen elsewhere, such as southeastern Australia. Finally, Kozhevnikov and Kozhevnikova[28] listed only 21 woody species (3%) among an overall list of 676 nonnative species of the Russian Far East, and only five from ENA, including *Acer negundo, R. pseudo-acacia, Fraxinus pennsylvanica, Populus deltoides,* and *Parthenocissus inserta (vitacea).*

In Europe, the restriction of invaders from ENA to particular habitats appears less extreme, although still biased toward the Asteraceae. One of the best characterized nonnative European floras is that of the Czech Republic,[29] which includes 1385 nonnative taxa, 90 of which are classified as invasive. The 30 invaders from ENA in the Czech Republic are a nearly even mix of annuals, herbaceous perennials, and woody species, with half of the herbaceous species represented by the Asteraceae; indeed, ENA accounts for about half of the invasive asteraceous species in the Czech flora, including many of the above meadow species that are invasive in East Asia. ENA woody species invasive in the Czech Republic also include most of those invasive in East Asia, plus *Prunus serotina*, a high-impact species that has spread throughout much of Europe.[30] Woody invaders overall, however, make up a relatively minor component of the nonnative Czech flora (17 species), and are represented by only one East Asian species (*Ailanthus altissima*). The large pool of East Asian woody invaders in ENA, particularly forest shrubs, appears to be absent from Europe, although other recent nonnative species inventories suggest their influence may be growing (e.g., in Italy,[31] where ENA composites are similarly invasive). Nevertheless, the biogeography of forest invasions in Europe and ENA seems distinct, with an Asian bias in ENA that is absent in Europe.

Development of Northern Hemisphere floras from the paleorecord

The apparent bias in modern biotic exchanges between Northern Hemisphere floras, particularly at the habitat level, invites speculation as to whether such biases can be linked to the contrasting evolutionary histories of the floras of each region. Until the mid to late Miocene (around 16–5 million years ago (Ma)), these regions and most of the Northern Hemisphere were united in a warm-temperate evergreen broadleaf and mixed forest—the Arctotertiary Geoflora[32]—with regions sharing strong climatic and taxonomic affinities.[33,34] Plant lineages across regions were shared through repeated colonization across both Beringia (until 5.5 Ma) and the North Atlantic Land Bridge (until around 15 Ma).[35] The Geoflora was progressively broken up by mid-continent aridification and cooling in the polar regions, with savanna and grassland expanding into the interior of Asia and North America between 12 and 5 Ma.[35,36] By the start of the Pliocene, evergreen broadleaf forests resembling those of the present southeastern United States were greatly contracted in ENA but remained well distributed in Europe and East Asia, leading to greater taxonomic similarity between the Old World regions than between them and ENA.[37] During the Pliocene, ENA was instead dominated by open habitats (savanna, grassland, shrubland, parklands) increasing in openness toward the continental interior, with coniferous forests in the Southeast, a small area of mixed forest in the mid-Atlantic region, and temperate deciduous forest confined to high latitudes.[37]

During the Pleistocene, ENA experienced repeated climate fluctuations that were harsher and more rapid than those of Europe or East Asia,[38,39] potentially exacerbating the open quality of ENA habitat types. By the Last Glacial Maximum (about 18,000 years ago), Adams and Faure[40] suggested that ENA below the glacial boundary was dominated by open conifer forests well into the continental interior and down to nearly the southern coast.[41–43] In contrast, Europe was largely steppe tundra and East Asia a mix of tundra, grassland, open temperate forest, and mesic broadleaf forest further south.[40,42,44–46] By 9000 years ago, however, most of ENA, Europe, and temperate East Asia had converged on closed canopy forests, with a more tropical element in East Asia that remains today.

Broadly interpreted, reconstructions of vegetation changes across the Northern Hemisphere since the time of a united geoflora in the late Miocene suggest major differences in the recent evolutionary history of modern ENA, European, and East Asian floras.[33] These can be summarized as (1) greater divergence toward open woodland or parkland vegetation in ENA during the Pliocene compared to the forests of Europe and East Asia,[37] which was likely maintained through multiple glaciation episodes in the Pleistocene;[40] (2) greater similarity of European vegetation to that of East Asia over the Pliocene, followed by large Pleistocene climatic disruptions across Europe that greatly expanded the extent of tundra at the expense of closed forests;[42] and (3) relative continuity of forest cover across temperate and subtropical latitudes in East Asia throughout the recent Cenozoic, with a strong connection to tropical forests throughout.[34,47,48]

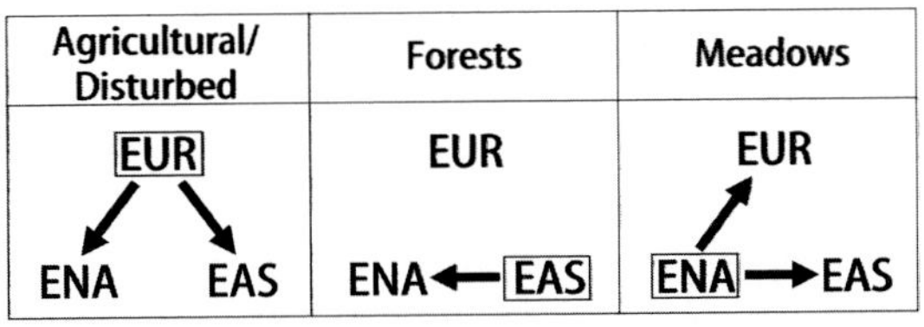

Figure 2. Apparent directionality of plant invasions for three habitat types between Europe (EUR), Eastern North America (ENA), and East/Far East Asia (EAS). Agricultural/disturbed habitats include cultivated fields, pastures, roadsides, lawns, and other systems of frequent disturbance; forests include predominantly shaded habitats during the growing season; and meadows include persistent, relatively undisturbed open habitats of rich substrate.

The case for preadaptation in the modern exchange of Northern Hemisphere floras

If a species is successful in an introduced ecosystem because it is a more effective competitor for scarce resources, and its competitive advantage stems from morphological or physiological novelties that evolved in its home range but not in the native flora, then that species can be said to be preadapted,[5] and the mechanism of invasion one of preadaptation or evolutionary innovation.[49] Conifers from Western North America, for example, are invasive at high elevations in the Southern Hemisphere because the Gondwanan flora is thought to be depauperate in frost-adapted trees due to its isolated history and low historical incidence of cold environments.[50] In the same way, one can ask whether the apparent bias in habitat exchanges across plant communities in the Northern Hemisphere may stem from differences in the evolutionary histories of their respective floras. I have summarized the major patterns of invasiveness by plants in each region and habitat (Fig. 2), and further outlined broad differences in the evolutionary history of each flora. Here, I tie these together and speculate whether a case can be made for preadaptation as a significant driver of Northern Hemisphere plant invasions. These observations are merely hypotheses; additional experimental work, such as home-and-away field comparisons of the competitive abilities and resource-use efficiencies of natives and invaders, is required for substantive development of the preadaptation invasion framework.

Why do European species dominate disturbed habitats?

The least contentious argument to be made concerning invasion through preadaptation is the case of European species in ecosystems that have been heavily modified by agricultural practices that originated in the Near East some 10,000 years ago and spread throughout Europe over the next few millennia.[19,51] Indeed, the wholesale movement of an entire ecosystem, including forest clearance for annual cereal cultivation and pasture as Europeans colonized much of the temperate world from the 16th century onward, is a dominant theme of environmental history.[52,53] In contrast to Europe, where cultivation-based landscapes have been common for several millennia, deforested landscapes of regular disturbance through tilling, grazing, and mowing have only been present in the Eastern United States for a few centuries.[54] It should come as no surprise, then, that few ENA natives have the innate biology to compete with Mediterranean and central European species in what are, for ENA, novel ecosystems: pastures, roadsides, lawns, cultivated ground, etc.,[8,10,19] just as is the case for other such transformed regions (e.g., annual invaders in the New Zealand flora).[5] This was apparent to even the earliest botanists in ENA.[9] Particularly interesting is the apparent lack of quality pasture forage in ENA, which may be due in part to a relatively short growing season for many native grasses and forbs compared to many Eurasian species[12] and the lack of dominant native grazers in ENA, at least during the Holocene.[55] Botanists and plant ecologists in the Eastern United States are familiar with the dearth of native species in chronically disturbed places; this seems more a comment on our cultural history than something in need of ecological explanation.

Why do East Asian and Russian Far East species dominate ENA forested habitats?

The large bias of species introduced from East Asia as invaders of forests of the Eastern United States must stem in part from the large number of species that have been introduced ornamentally from China, Korea, and Japan.[13] However, even given this bias, plants from this region are far more likely to be invasive than woody plants from other regions, including Europe.[8] The majority of the species in question are shrubs and vines, typically fully to partly shade tolerant,[56] and with only a few exceptions present

at the genus level in the native ENA flora.[57] Why should East Asian species be particularly good at invading ENA forests?

Fridley[57] suggested that the long evolutionary history of diverse lineages in closed forests in East Asia may have led to more effective strategies for resource capture in an understory environment—that is, the invaders are inherently better understory competitors than most of the native ENA species. One strategy in particular—the maintenance of carbon acquisition well after canopy leaf fall in autumn—was shown to be systematically greater for nonnative forest invaders compared to native species, even within the same genus.[57] The autumn advantage was not specific to East Asian species, however, and other studies have found physiological and demographic advantages for both Asian and European species versus native ENA species.[58,59] In a specific comparison of the functional properties of East Asian and ENA species, Heberling and Fridley[60] found advantages for East Asian species for some, but not all, metabolic cost–benefit relationships. Ongoing research continues to find important physiological differences between native and invasive forest species in ENA (Fridley, Heberling, Jo, unpublished data), but such advantages do not appear to be restricted to East Asian species.

Importantly, any preadaptation argument for the East Asian bias in woody invaders in ENA must further explain why, thus far, these species appear to have limited invasion potential in the forests of Europe, despite the fact that European forests are clearly invasible.[61] Moreover, European forests typically have lower woody diversity than those of similar climates in ENA,[62] and East Asian species may be less likely to encounter ecologically similar relatives in Europe than in ENA.[57] One possibility is that invaders benefit more from the presence of earthworms in ENA, which have coevolved with both East Asian and European plant lineages but were largely absent from much of ENA until recent introductions.[63] In this case, East Asian species may be less competitively advantaged against European species that also experienced a long association with the rhizospheric effects of earthworms.[64]

Why do ENA composites spread throughout Northern Hemisphere meadows?

The largest contribution of the ENA flora to the invasive floras of Europe and Asia is its composite

(Asteraceae) flora, including perennial species common to seminatural meadows of open sites (e.g., *Solidago altissima*) and partly shaded, mesic locations (e.g., *Ageratina altissima, Rudbeckia laciniata, Symphyotrichum lanceolatum*), as well as erect annuals capable of creating large monocultures after disturbance (e.g., *Ambrosia artemisiifolia, Conyza canadensis, Erigeron annuus, Helianthus annuus*). It is notable that nearly all the species in question are tall, fast-growing, drought-intolerant, and often found in relatively fertile locations, particularly bottomlands, with the annuals favored in more open sites.[65,66] These are not deep forest species,[67] nor do they seem particularly tolerant to defoliation or low nutrient availability. In a classification of plant strategies, they fit the competitor or competitive ruderal mode, common to plants in temperate ecosystems of occasional disturbance and abundant resources, including meadows and riparian areas.[68] Such ecosystems tend to be on relatively nutrient-rich, geologically young substrate.[69]

It is intriguing that these are likely the conditions commonly experienced by the ENA flora over the past several million years, inhabiting a region that may have been dominated by open woodlands since the Pliocene[37] and subject to repeated soil nutrient enrichment from Pleistocene glaciations, at least over its northern half.[70] Given the richness of native ENA taxa, including composites, that dominate such habitats today, it is compelling to speculate where this group of species would have persisted in ENA during recent evolutionary time if ENA had instead been dominated by closed forests.[67] Marks[71] concluded that many of the same species in question—including *A. artemisiifolia, C.(Erigeron) canadensis*, and *S. altissima*—likely evolved in persistent open habitats rather than ephemeral forest gaps, in part because these species are generally less well-dispersed than those more typical of frequently disturbed habitats today. Although difficult to test formally due to the speculative nature of reconstructing past habitat conditions, if ENA experienced a greater frequency of meadow- or parkland-type vegetation than Europe or East Asia over the past several million years, it could explain both the bias in Asteraceae in mesic-temperate North America and the tendency of some of these species to invade similar habitats across the Northern Hemisphere.

Conclusion

I have drawn a portrait of modern floristic exchanges between regions of the Northern Hemisphere that emphasizes geographic directionality and habitat specificity, and have suggested that these patterns are partly the result of evolutionary processes that have occurred over millions of years in response to large-scale shifts in climate, soil conditions, and disturbance regimes. Although my intention is to elevate the status of preadaptation as a mechanism of biological invasion, this does not deny the importance of other, more proximate ecological mechanisms, including biotic resistance (whether more diverse ecosystems are less invasible) or enemy release (whether species become invasive by escaping their native predators or pathogens), which could be contributing factors in certain cases. It is further important to acknowledge the critical role played by local cultural factors, particularly land use and introduction effort, in the spread of invasive species. However, I note that mechanisms that only invoke local processes, without considering a larger biogeographic or evolutionary context, cannot explain large biases in the directionality of invasions between global biotic regions (in addition to those listed here, consider imbalances in plant exchanges to and from oceanic islands, New Zealand, between the Northern and Southern Hemispheres [as even noted by Darwin[72]], and between the major Mediterranean regions, among many such examples). Vastly different rates of biotic exchange between regions, even after accounting for differences in introduction rates, strongly point to invasion mechanisms that are evolutionary in nature, reflecting a global imbalance in the extent to which certain plant and animal lineages are adapted to modern ecosystems. Insights into how adaptive evolution has shaped the floristic composition of past and present habitats are likely to be key tools for predicting how the biosphere itself will evolve in the coming decades.

Acknowledgments

I am grateful to John Wiersema for provision of the GRIN Taxonomy dataset, Hyo Hue Mi Lee, Kang-Hyun Cho, and Insu Jo for provision of the Korean nonnative plants database, an anonymous reviewer for manuscript comments, and Andrey Kozhevnikov, Mason Heberling, Guillaume Decocq, and Jonathan Adams for discussion of Northern Hemisphere invasion patterns and the paleorecord. I gratefully acknowledge the generous support of Len Blavatnik, the Blavatnik Family Foundation, and the New York Academy of Sciences.

Conflicts of interest

The author has no conflicts of interest.

References

1. Elton, C. 1958. *The Ecology of Invasions by Animals and Plants*. Chicago: University of Chicago Press.

2. Richardson, D.M. (Ed). 2011. *Fifty Years of Invasion Ecology: the Legacy of Charles Elton*. London: Blackwell.

3. Behrensmeyer, A.K., J.D. Damuth, W.A. DiMichele, *et al.* 1992. *Terrestrial Ecosystems through Time: Evolutionary Paleoecology of Terrestrial Plants and Animals*. Chicago: University of Chicago Press.

4. Vermeij, G.J. 1991. When biotas meet: understanding biotic interchange. *Science* **253**: 1099–1104.

5. Mack, R.N. 2003. Phylogenetic constraint, absent life forms, and preadapted alien plants: a prescription for biological invasions. *Int. J. Plant Sci.* **164**: S185-S196.

6. Vermeij, G.J. 2005. Invasion as expectation: a historical fact of life. In *Species Invasions: Insights into Ecology, Evolution, and Biogeography*. D.F. Sax, S.D. Gaines & J.J. Stachowicz, Eds.: 315–339. Sunderland, MA: Sinauer.

7. Vermeij, G.J. 1996. An agenda for invasion biology. *Biol. Cons.* **78**: 3–9.

8. Fridley, J.D. 2008. Of Asian forests and European fields: Eastern U.S. plant invasions in a global floristic context. *PLoS One* **3**: e3630.

9. Mehrhoff, L.J. 2000. Immigration and expansion of the New England flora. *Rhodora* **102**: 280–298.

10. Boufford, D.E. 2001. Introduced species and the 21st century floras. *J. Jap. Bot.* **76**: 245–262.

11. Mack, R.N. 2003b. Plant naturalizations and invasions in the Eastern United States: 1634–1860. *Ann. Miss. Bot. Gar.* **90**: 77–90.

12. Anderson, E. 1952. *Plants, Man and Life*. New York: Dover.

13. Rehder, A. 1936. On the history of the introduction of woody plants in North America. *Nat. Hort. Mag.* 245–257.

14. USDA, NRCS. 2007. The PLANTS Database. National Plant Data Center, Baton Rouge, LA. http://plants.usda.gov. Accessed April 11, 2007.

15. USDA, ARS, National Genetic Resources Program. 2012. *Germplasm Resources Information Network (GRIN)* (Online Database). Beltsville, Maryland: National Germplasm Resources Laboratory. URL: http://www.ars-grin.gov/cgi-bin/npgs/html/index.pl?language = en

16. Braun, E.L. 1950. *Deciduous Forests of Eastern North America*. Philadelphia, PA: Blakiston.

17. Takhtajan, A. 1986. *Floristic Regions of the World*. Berkeley: University of California Press.

18. Hollis, S. & R.K. Brummitt. 1992. *World Geographical Scheme for Recording Plant Distributions*. Plant Taxonomic Database Standards No. 2. Version 1.0. Published for the International Working Group on Taxonomic Databases for Plant Sciences (TDWG) by the Hunt Institute for Botanical Documentation, Pittsburgh: Carnegie Mellon University.

19. di Castri, F. 1989. History of biological invasions with special emphasis on the Old World. In *Biological Invasions: a Global Perspective*. J.A. Drake, H.A. Mooney, F. di Castri, R.H. Groves, F.J. Kruger, M. Rejmánek, & M. Williamson, Eds.: 1–30. Chichester, UK: John Wiley & Sons.

20. Weber, E., S.-G. Sun & B. Li. 2008. Invasive alien plants in China: diversity and ecological insights. *Biol. Inv.* **10**: 1411–1429.

21. Park, S. 2009. *Naturalized Plants of Korea*. Seoul, South Korea: Ilchokak Seoul.

22. Korea Forest Service. 2010. *Korean Plant Names Index*. http://www.nature.go.kr/kpni/general/Prgb01/Prgb1_1.jsp

23. Lee, T. B. 1993. *Illustrated Flora of Korea*. Seoul, South Korea: Hyang Mun Sa

24. Park, S., Y. Lee, S. Jung, S. Jung & S. Oh. 2012. *Field Guide: Naturalized Plants of Korea*. Korea: National Arboretum.

25. Choung, Y., W. T. Lee, K.-H. Cho, K. Y. Joo, B. M. Min, J.-O. Hyun & K. S. Lee. 2012. *Categorizing Vascular Plant Species Occurring in Wetland Ecosystems of the Korean Peninsula*. Chuncheon, Korea: Center for Aquatic Ecosystem Restoration.

26. Miyawaki, S. & I. Washitani. 2004. Invasive alien plant species in riparian areas of Japan: the contribution of agricultural weeds, revegetation species, and aquacultural species. *Glob. Env. Res.* **8**: 89–101.

27. Auld, B., H. Morita, T. Nishida, M. Ito & P. Michael. 2003. Shared exotica: plant invasions of Japan and south eastern Australia. *Cunninghamia* **8**: 147–152.

28. Kozhevnikov A.E. & Z.V. Kozhevnikova. 2011. Alien species plant complex as a component of the Far East of Russia natural flora: diversity and regional changes of taxonomical structure. *Dalnauka* **58**: 5–36.

29. Pyšek, P., J. Sádlo & B. Mandák. 2002. Catalogue of alien plants of the Czech Republic. *Preslia* **74**: 97–186.

30. Starfinger, U. 1997. Introduction and naturalization of *Prunus serotina* in central Europe. In *Plant Invasions: Studies from North America and Europe*. J.H. Brock, M. Wade, P. Pyšek, & D. Green, Eds.: 161–171. Leiden, the Netherlands: Backhuys.

31. Celesti-Grapow, L., A. Alessandrini, P.V. Arrigoni, *et al.* 2009. Inventory of the non-native flora of Italy. *Plant Biosys.* **143**: 386–430.

32. Wen, J. 1999. Evolution of Eastern Asian and Eastern North American disjunct distributions in flowering plants. *Annu. Rev. Ecol. Syst.* **30**: 421–455.

33. Manchester, S.R. 1999. Biogeographical relationships of North American Tertiary floras. *Ann. Miss. Bot. Gar.* **86**: 472–522.

34. Donoghue, M.J. & S.A. Smith. 2004. Patterns in the assembly of temperate forests around the Northern Hemisphere. *Phil. Trans. R. Soc. Lond. B* **359**: 1633–1644.

35. Milne, R.I. 2006. Northern Hemisphere plant disjunctions: a window on Tertiary land bridges and climate change? *Ann. Bot.* **98**: 465–472.

36. Pound, M.J., A.M. Haywood, U. Salzmann & J.B. Riding. 2012. Global vegetation dynamics and latitudinal temperature gradients during the Mid to Late Miocene (15.97–5.33 Ma). *Earth-Sci. Rev.* **112**: 1–22.

37. Salzmann, U., M. Williams, A.M. Haywood, *et al.* 2011. Climate and environment of a Pliocene warm world. *Palaegeogr. Palaeoclim. Palaeoecol.* **309:** 1–8.

38. Ehlers, J. & P.L. Gibbard. 2007. The extent and chronology of Cenozoic Global Glaciation. *Quaternary Int.* **164–165:** 6–20.

39. Sandel, B., L. Arge, B. Dalsbaard, *et al.* 2011. The influence of Late Quaternary climate-change velocity on species endemism. *Science* **334:** 660–664.

40. Adams J.M. & H. Faure. 1997. Preliminary vegetation maps of the world since the last glacial maximum: an aid to archaeological understanding. *J. Archaeol. Sci.* **24:** 623–647.

41. Ray, N. & J.M. Adams. 2001. A GIS-based vegetation map of the world at the last glacial maximum (25,000–15,000 BP). *Internet Archaeol.* **11:** 3.

42. Prentice, I.C., D. Jolly & BIOME 6000 Participants. 2000. Mid-Holocene and glacial-maximum vegetation geography of the northern continents and Africa. *J. Biogeogr.* **27:** 507–519.

43. Harrison, S.P. & I.C. Prentice. 2003. Climate and CO_2 controls on global vegetation distribution at the last glacial maximum: analysis based on palaeovegetation data, biome modelling and palaeoclimate simulations. *Glob. Change Biol.* **9:** 983–1004.

44. Liew, P.M., C.M. Kuo, S.Y. Huang & M.H. Tseng. 1998. Vegetation change and terrestrial carbon storage in eastern Asia during the Last Glacial Maximum as indicated by a new pollen record from central Taiwan. *Glob. Planet. Change* **16-17:** 85–94.

45. Takahara, H., S. Sugita, S.P. Harrison, *et al.* 2000. Pollen-based reconstructions of Japanese biomes at 0, 6000 and 18,000 ^{14}C yr BP. *J. Biogeogr.* **27:** 665–683.

46. Hope, G., A.P. Kershaw, S. van der Kaars, *et al.* 2004. History of vegetation and habitat change in the Austral-Asian region. *Quaternary Int.* **118-119:** 103–126.

47. Guo, Q., R.E. Ricklefs & M.L. Cody. 1998. Vascular plant diversity in eastern Asia and North America: historical and ecological explanations. *Bot. J. Linn. Soc.* **128:** 123–136.

48. Xiang, Q.-Y., W.H. Zhang, R.E. Ricklefs, *et al.* 2004. Regional differences in rates of plant speciation and molecular evolution: a comparison between Eastern Asia and Eastern North America. *Evolution* **58:** 2175–2184.

49. Fridley, J.D. 2011. Biodiversity as a bulwark against invasion: conceptual threads since Elton. In *Fifty Years of Invasion Ecology: the Legacy of Charles Elton.* D.M. Richardson, Ed.: 121–130. London: Blackwell.

50. Körner, C. & J. Paulsen. 2004. A world-wide study of high altitude treeline temperatures. *J. Biogeogr.* **31:** 713–732.

51. Pinhasi, R., J. Fort & A.J. Ammerman. 2005. Tracing the origin and spread of agriculture in Europe. *PLoS Biol.* **3:** e410.

52. Crosby, A.W. 1986. *Ecological Imperialism: the Biological Expansion of Europe, 900–1900.* Cambridge: Cambridge University Press.

53. Mann, C. 2011. *1493: Uncovering the New World Columbus Created.* New York: Vintage.

54. Mönkkönen, M. & D.A. Welsh. 1994. A biogeographical hypothesis on the effects of human caused landscape changes on the forest bird communities of Europe and North America. *Ann. Zool. Fennici.* **31:** 61–70.

55. Lorenzen, E.D., D. Nogués-Bravo, L. Orlando, *et al.* 2011. Species-specific responses of Late Quaternary megafauna to climate and humans. *Nature* **479:** 359–364.

56. Martin, P.H., C.D. Canham & P.L. Marks. 2009. Why forests appear resistant to exotic plant invasions: intentional introductions, stand dynamics, and the role of shade tolerance. *Front. Ecol. Environ.* **7:** 142–149.

57. Fridley, J.D. 2012. Extended leaf phenology and the autumn niche in deciduous forest invasions. *Nature* **485:** 359–362

58. Harrington, R. A., B. J. Brown & P. B. Reich. 1989. Ecophysiology of exotic and native shrubs in southern Wisconsin: 1. Relationship of leaf characteristics, resource availability, and phenology to seasonal patterns of carbon gain. *Oecologia* **80:** 356–367.

59. Martin, P.H., C.D. Canham & R.K. Kobe. 2010. Divergence from the growth-survival trade-off and extreme high growth rates drive patterns of exotic tree invasions in closed-canopy forests. *J. Ecol.* **98:** 778–789.

60. Heberling, J.M. & J.D. Fridley. 2012. Biogeographic constraints on the worldwide leaf economics spectrum. *Glob. Ecol. Biogeogr.* **21:** 1137–1146.

61. Chabrerie, O., K. Verheyen, R. Saguez & G. Decocq. 2007. Disentangling relationships between habitat conditions, disturbance history, plant diversity, and American black cherry (*Prunus serotina* Ehrh.) invasion in a European temperate forest. *Div. Distrib.* **14:** 204–212.

62. Manthey, M., J.D. Fridley & R. K. Peet. 2011. Niche expansion after competitor extinction? A comparative assessment of habitat generalists and specialists in the tree floras of southeastern North America and southeastern Europe. *J. Biogeogr.* **38:** 840–853.

63. Nuzzo, V.A., J.C. Maerz & B. Blossey. 2009. Earthworm invasion as the driving force behind plant invasion and community change in Northeastern North American forests. *Conserv. Biol.* **23:** 966–974.

64. Frelich, L.E., C.M. Hale, S. Scheu, *et al.* 2006. Earthworm invasion into previously earthworm-free temperate and boreal forests. *Biol. Invas.* **8:** 1235–1245.

65. Gleason, H.A. & A. Cronquist. 1991. *Manual of Vascular Plants of Northeastern United States and Adjacent Canada.* New York, Bronx: Botanical Garden.

66. Weakley, A.S. 2012. *Flora of the Southern and Mid-Atlantic States.* September 2012 version. UNC Herbarium, North Carolina Botanical Garden, University of North Carolina at Chapel Hill.

67. Gleason, H.A. & A. Cronquist. 1964. *The Natural Geography of Plants.* New York: Columbia University Press.

68. Grime, J.P. 1977. Evidence for the existence of three primary strategies in plants and its relevance to ecological and evolutionary theory. *Am. Nat.* **111:** 1169–1194.

69. Jenny, H. 1941. *Factors of Soil Formation: a System of Quantitative Pedology.* New York: Dover.

70. von Englen, O.D. 1914. Effects of continental glaciation on agriculture. Part I. *Bull. Am. Geogr. Soc.* **46:** 241–264.

71. Marks, P.L. 1983. On the origin of field plants of the Northeastern United States. *Am. Nat.* **122:** 210–228.

72. Darwin, C. 1859. *On the Origin of Species.* New York: Facsimile reproduction of the First Edition by Atheneum Press.

Ann. N.Y. Acad. Sci. ISSN 0077-8923

ANNALS OF THE NEW YORK ACADEMY OF SCIENCES

Issue: *Blavatnik Awards for Young Scientists 2012*

Functional genomics lead to new therapies in follicular lymphoma

Elisa Oricchio and Hans-Guido Wendel

Cancer Biology & Genetics Program, Memorial Sloan-Kettering Cancer Center, New York, New York

Address for correspondence: Elisa Oricchio, Memorial Sloan-Kettering Cancer Center, Cancer Biology & Genetics Program, 1275 York Ave, Box 337, New York, NY 10065. oriccche@mskcc.org

Recent technological advances allow analysis of genomic changes in cancer in unprecedented detail. The next challenge is to prioritize the multitude of genetic aberrations found and identify therapeutic opportunities. We recently completed a study that illustrates the use of unbiased genetic screens and murine cancer models to find therapeutic targets among complex genomic data. We genetically dissected the common deletion of chromosome 6q and identified the ephrin receptor A7 (*EPHA7*) as a tumor suppressor in lymphoma. Notably, *EPHA7* encodes a soluble splice variant that acts as an extrinsic tumor suppressor. Accordingly, we developed an antibody-based strategy to specifically deliver *EPHA7* back to tumors that have lost this gene. Recent sequencing studies have implicated *EPHA7* in lung cancer and other tumors, suggesting a broader therapeutic potential for antibody-mediated delivery of this tumor suppressor for cancer therapy. Together, our comprehensive approach provides new insights into cancer biology and may directly lead to the development of new cancer therapies.

Keywords: genomic data; mouse model; therapies

Introduction

Cancer is a complex disease characterized by myriad genetic alterations that affect many pathways and processes.[1–5] The cancer genome is altered at the nucleotide, chromosomal, and epigenetic levels. However, only a small number of mutations are required for malignant transformation; for example, in acute myeloid leukemia only two to three mutations are sufficient to cause leukemia,[6] implying that, while many of the changes identified may contribute to the malignant phenotype, they may be dispensable and therefore less likely to be important for cancer therapy. Identifying critical changes is complicated by significant inter- and intratumoral heterogeneity, and despite advanced statistical analyses, the descriptive data alone may not pinpoint the critical lesions that are amenable to therapy. Functional assays are needed to define the biological impact of genetic alterations and to understand their requirement for tumor development and also maintenance of the malignant cells. Therefore, this review emphasizes complementary approaches that include large-scale genomic data and genetic experiments to functionally annotate cancer genome data.

Cancer genomics and potential therapies

Genetic screens are unbiased experimental tools to simultaneously assess the biological activities of large numbers of genes. Well-designed screens are excellent assays to investigate complex genomic data, and the unbiased nature of this approach ensures that new lesions can be identified. Genetic screens are performed in different laboratories using different technologies. These include gain- and loss-of-function studies, both *in vitro* and *in vivo*, which may be performed on a one-by-one basis or as massively parallel experiments with pooled libraries.[7–9] Our lab performs large-scale, pooled, short-hairpin RNA (shRNA) library screens, an ideally suited approach for prioritizing gene lists. For example, screening can be performed to identify functional candidates in large genomic regions affected by copy number alterations, or to study genes that have broad-acting biological function, such as

doi: 10.1111/nyas.12120

Ann. N.Y. Acad. Sci. 1293 (2013) 18–24 © 2013 New York Academy of Sciences.

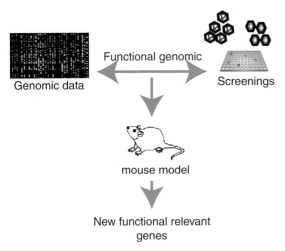

Figure 1. A combination of powerful tools to functionally annotate genomic data. Analysis of cancer genomic data provides a list of relevant alterations for functional validation. Genetic screens are versatile tools to simultaneously test many genes for subsequent validation in a physiological context.

those involved in epigenetic regulation, transcription factors, or microRNAs.[1,9] The choice of screening experiment is a key consideration and the experimental system will determine the biological activity that can be identified. Clearly, an *in vitro* screen for transforming activities can only capture a subset of the biological processes involved in malignant transformation and will be blind to factors enabling immune escape—these would be considered false negative as they will not be found. Screens are further complicated by false positive "hits." Although repeated screens will eliminate most of them, some hits may have true activity in the screen system but lack significance *in vivo* or in the human disease. The latter may only be revealed upon individual validation in a genetically and pathologically accurate *in vivo* cancer model. In short, unbiased genetic screens are highly versatile tools to filter large-scale genomic data for subsequent *in vivo* validation studies.[1,10,11]

Murine models of cancer enable direct functional studies of genetic alterations seen in human tumors (Fig. 1). In particular, mosaic models enable rapid assessment of genetic changes in a physiological context,[11–13] an example of which is based on the adoptive transfer of virally transduced bone marrow cells into wild-type syngeneic recipient animals. This approach can be used to model several hematopoietic malignancies *in vivo*. Impor-

tantly, the ease of genetic modification avoids the time-consuming generation of mutant or transgenic animals for each allele to be studied. Our approach is based on the retroviral modification and transplantation of hematopoietic progenitors cells (HPCs) and allows rapid testing of several genes of interest.[9,13,14] Moreover, using HPC derived from animals with different genetic backgrounds allows modeling of different forms of lymphoma, including the EμMyc murine model of Burkitt-like lymphoma,[15] or the IμHABCL6[16] and Eμ-*CyclinD1*,[17] used to recapitulate the genetics of diffuse large B cell lymphoma (DLBCL) and mantle cell lymphoma, respectively. Most recently, we developed an adoptive transfer model based on vavBcl2 transgenic mice[12] that develop a disease that resembles follicular lymphoma in genetics and pathological appearance, including evidence of somatic hypermutation and germinal center origin.[12] For the first time, this model enables the study of nontransformed follicular lymphoma (FL) to examine genetic and cellular interactions involved in lymphomagenesis, treatment response, and malignant progression. In conjunction with human genomic data and unbiased genetic screens, these murine models can provide direct evidence of pathogenic drivers and genetic requirements in human cancer.

We applied a functional genomics strategy to identify driver mutations in FL,[1] the most common indolent form of nonHodgkin lymphoma. The cytogenetic hallmark of FLs is the chromosomal translocation t(14:18)(q32;q21) that results in constitutive expression of the antiapoptotic gene *Bcl2*.[18] However, additional genetic lesions are required to induce lymphomagenesis or disease progression, which occurs in 50% of cases.[19,20] For example, c-MYC amplification, loss of p53, and deletions of chromosome 6 have been linked with disease progression and reduced survival.[21–24] Overall, the median survival of patients with FL is approximately 8–10 years and is limited in 50% of cases by the transformation into aggressive B cell lymphoma; the remainder of patients show a pattern of incessant disease relapse following chemotherapy that eventually outruns the patients marrow reserve, leading to transfusion dependence and recurrent infections. The inclusion of the anti-CD20 antibody (Rituxan®) in the treatment of FL 15 years ago has been the last major improvement and has significantly impacted patient outcomes.[25,26] Despite this advance, bone marrow transplantation remains the only curative option in suitable patients. It has been argued that survival times in this cancer are long, and that, with an elderly patient population, further improvements may be hard to accomplish. However, we think that new insights into the pathogenesis of FL can lead to more effective and potentially less toxic alternatives to the current combinations of chemotherapy and Rituxan.

Scientifically, FL has not received much attention compared to aggressive lymphomas. In part, this is due to a lack of nontransformed FL cell lines and adequate animal models. Accordingly, most studies have sought to correlate patient features and individual genetic changes with the outcome under available therapy, and to inform doctors' decisions on whether and when to initiate treatment.[27,28] Only recent advances in sequencing technology have provided new insights into the genetic make-up of FL, including the unexpected finding that the most frequently mutated genes in FL are involved in epigenetic control of gene expression (e.g., *EZH2*, *MLL2*, *MEF2B*, and *CREBBP*).[29,30] Among other findings, recurrent mutations in *B2M* and *CD58* implicate immune escape, and activation of B cell receptor signals that are most likely endogenous or mutational also contribute to FL pathogenesis.[31,32] In addition to these recurrent somatic mutations, genomic analyses reveal recurrent chromosomal gains and losses, including deletions of chromosome 6q, which occur in approximately 25% of FLs and are associated with poor prognosis.[1,33] Typically, 6q deletions are large and hemizygous, and thus the genetic target(s) of chromosome 6q loss has long been an enigma. Recently, several tumor suppressor genes were identified for diffuse large B cell lymphoma that are encoded on chromosome 6q, including *TNFAIP3*[34,35] and *PRDM1/BLIMP*.[36] The pattern and size of 6q deletions indicate the existence of multiple tumor suppressor activities in this region.

To search for genes that may contribute to FL development, we designed a retroviral shRNA library against all genes encoded within regions of loss on chromosome 6q, and each gene was targeted by two to five different short hairpins. An unbiased genetic screen was performed for cooperation with *BCL2* in B lymphocytes *in vitro* to identify shRNAs that provide proliferative advantage. The screening identifies several interesting candidates that were validated individually. *TNFAIP3* was

immediately confirmed as a tumor suppressor in lymphoma, and *EPHA7* was identified as a new candidate gene. A short splicing variant of *EPHA7* that encodes a soluble protein is normally expressed in B lymphocytes and is lost in up to 75% of FLs.[37] We used vavP-Bcl2 chimeric mouse model of FL to test and validate *EPHA7* activity in a specific tumor context, and readily confirmed that a loss of *EPHA7* promotes FL development, and conversely, its reexpression slows tumor growth *in vivo*.

Ephrin receptors are a large family of receptor tyrosine kinases involved in cell–cell signaling, and alterations in these genes have been implicated in solid cancers.[38,39] Specifically, ephrin receptors are activated by binding ephrins (ligands) that are expressed on the cell surface. These ligand–receptor interactions stimulate formation of duplex and higher order ephrin-receptor clusters, which are involved in cell signals. Notably, a cellular signal emanates in both directions from ligand and receptor, and changes cell behavior. This signaling pathway has been implicated in mechanisms controlling brain size, axon guidance, and retina formation.[40,41] Within the family of ephrin receptors, *EPHA7* is notable for the existence of a short splice variant that encodes a truncated protein (EPHA7TR). The truncated EPHA7TR can be shed from the cell surface and acts as a dominant inhibitor of the full-length receptor that forms inactive heterodimers with full-length ephrin receptors. This surprising mode of action for *EPHA7* has previously been implicated in closure of the neural tube during embryonal development.[42] Although *EPHA7* is highly expressed in lymphoid tissues, its physiological role in B cell development or activation has not been explored. Importantly, the cell-extrinsic mode of action suggests that exogenous administration may be able to restore *EPHA7* function.

In follicular lymphoma, *EPHA7* is deleted and silenced by promoter methylation in over 70% of the tumors; this percentage may be even higher in aggressive lymphomas. *EPHA7* acts in the pathogenesis of FL by binding and blocking the activity of the EPHA2 receptor, which then inactivates ERK and SRC kinase pathways. Notably, treatment of cultured or xenografted lymphoma cells with *EPHA7* inhibits ERK and SRC activity and causes cell death, suggesting a potential therapeutic application for *EPHA7* in lymphoma therapy. Moreover, in analogy with the concept of oncogene dependence,[43] it

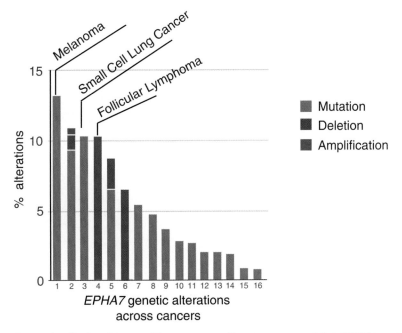

Figure 2. *EPHA7* is mutationally altered in several human cancers. Recent papers report that *EPHA7* is mutated or deleted in different cancers including melanoma, small cell lung cancer, and lymphoma, implying a potential application of *EPHA7*-based therapies for these cancers.

has been proposed that tumor cells are exquisitely sensitive to the restoration of tumor suppressor genes. Indeed, studies in genetically engineered mice, where a tumor suppressor can be reactivated, indicate powerful antitumor activity.[44,45] For example, restoring p53 function has been demonstrated to have a powerful, tissue-specific mechanism that suppresses tumorigenesis,[44] supporting the idea that pharmacological reactivation of tumor suppressor genes is a therapeutic possibility. However, most tumor suppressors act cell intrinsically, and are mutationally lost and not amenable to exogenous restoration. Exogenous administration of *EPHA7* may represent a rare instance where tumor suppressive activity can be restored.

We tested different routes for administering the soluble EPHA7 protein for lymphoma therapy. Although local administration was extremely powerful, this is not a satisfying therapy for a systemic disease-like lymphoma. Systemic intravenous administration showed antitumor activity and was well tolerated; however, it proved far less effective than the local application. Finally, we explored the possibility of using the anti-CD20 (Rituximab) antibody to deliver effective concentrations of *EPHA7*

to the lymphomas *in vivo*. This approach was highly effective and combined the intrinsic antitumor effect of the anti-CD20 antibody with the specific tumor suppressor effect of *EPHA7* that is able to shut down ERK and SRC signals in lymphoma cells. The fusion of *EPHA7* with the anti-CD20 antibody represents an improvement over anti-CD20 alone, and reflects the specific sensitivity of cancer cells toward the restoration of a tumor suppressor.

Recently, *EPHA7* has also been implicated in other cancers: its expression is lost in DLBCL, acute B lymphoblastic leukemia (B-ALL), sarcoma, colon, and prostate cancers.[46–49] Recently, deep sequence analyses revealed common mutations in *EPHA7* in small cell lung cancer, melanoma, head and neck carcinoma, and other cancers[50–55] (Fig. 2 and Table 1). Together, these data indicate an important role for ephrin signaling and *EPHA7* in a spectrum of human cancers. We are currently testing the potential of *EPHA7* administration to treat other tumor types.

Perspective

Cancer genomic studies improved cancer therapy. In the past few years, drugs targeting specific genetic alterations have been successfully developed.[56–59]

Table 1. Summary of reported *EPHA7* genetic alterations

	Cancer Type	Alterations	Frequency	Reference
1	Skin cutaneous melanoma	Mutations	13.2	Hodis *et al.*[52]
2	Lung adenocarcinoma	Mutations	9.3	Imielinksi *et al.*[53]
		Deletions	1.1	
		Amplifications	0.5	
3	Small cell lung	Mutations	10.3	Peifer *et al.*[50]
4	Follicular lymphoma	Deletions	11	Oricchio *et al.*[1]
5	Cell line encyclopedia	Mutations	6.5	Barretina *et al.*[49]
		Deletions	2.2	
6	Prostate adenocarcinoma	Deletions	6.6	Grasso *et al.*[54]
7	Head and neck	Mutations	5.4	Stransky *et al.*[51]
8	Small cell lung	Mutations	4.8	Rudin *et al.*[55]
9	Lung adenocarcinoma	Mutations	3.7	Ding *et al.*[63]
10	Colorectal	Mutations	2.8	Seshagiri *et al.*[64]
11	Medulloblastoma	Mutations	2.7	Robinson *et al.*[65]
12	Breast	Mutations	2	Stephens *et al.*[66]
13	Kidney	Mutations	2	Guo *et al.*[67]
14	Breast	Mutations	1.9	Banerji *et al.*[68]
15	Prostate adenocarcinoma	Mutations	0.9	Barbieri *et al.*[69]
16	Medulloblastoma	Mutations	0.8	Jones *et al.*[70]

Source: www.cbioportal.org

The success of these therapies is based, at least in part, on the continuous requirement of the initiating oncogenic signals for tumor maintenance. Recent sequencing studies have highlighted the important role of recurrent somatic mutations in individual genes. Somatic mutagenesis is not the only process that shapes the cancer genome. Additional changes such as gains and losses of chromosomal material or activation and silencing of complex transcriptional programs are likely just as important. However, the functional analysis of such large-scale alterations remains a major challenge. We described a highly efficient strategy to prioritize large gene sets altered in human tumors that is based on combination of unbiased genetic screens and the use of highly versatile animal models of cancers. In this manner, we can identify key driver genes within large regions of genomic loss that lend themselves to new therapies.

Chromosomal gains and losses are not precise mutagenic events like point mutations. However, the recurrent nature of these lesions indicates that they confer fitness advantage and contribute to tumor development or progression. Chromosomal aberrations likely affect multiple tumor suppres-

sor activities that may produce cooperative effects,[60] which is the case with the large 6q deletions seen in lymphoma. Notably, the restoration of one tumor suppressor (*EPHA7*) is sufficient to produce dramatic therapeutic effects. The imprecise nature of these lesions may also affect genes whose inactivation is not beneficial to the tumor cell,[61,62] potentially resulting in collateral sensitivity of tumor cells that have lost one gene copy toward pharmacological inhibitors.

Does the *EPHA7* tumor suppressor have therapeutic activity beyond lymphoma?

Recently, multiple genomic studies have indicated that *EPHA7* is a mutational target across multiple cancers (Fig. 2), opening the possibility to extend the therapeutic use of *EPHA7* to different tumors. Notably, *EPHA7* is one of very few soluble tumor suppressors that lend themselves to exogenous administration. We have been able to specifically deliver *EPHA7* to lymphomas by fusing *EPHA7* to the anti-CD20 antibody already used in lymphoma therapy. Similarly, antibodies have been used to deliver toxins or radioactive compounds to tumors cells.

The targeted delivery of a tumor suppressor has an important conceptual advantage over the targeted delivery of broad acting toxins, namely, cancer cells are especially sensitive to the restoration of a tumor suppressor that has been lost during development of the tumor, a phenomenon that has been called "tumor suppressor hypersensitivity,"[45] which adds to the specific antitumor effect of the fusion construct.

Acknowledgments

This work is supported by grants from the NCI (R01-CA142798-01), and a P30 supplemental award (HGW), the Leukemia Research Foundation (HGW), the Louis V. Gerstner Foundation (HGW), the WLBH Foundation (HGW), a grant from the American cancer Society (HGW), and Leukemia and Lymphoma Research Foundation Special Fellowship (EO).

Conflicts of interest

The authors declare no conflicts of interest.

References

1. Oricchio, E. *et al.* 2011. The Eph-receptor A7 is a soluble tumor suppressor for follicular lymphoma. *Cell* 147: 554–564.
2. Beroukhim, R. *et al.* 2010. The landscape of somatic copy-number alteration across human cancers. *Nature* 463: 899–905.
3. Segal, E., N. Friedman, D. Koller & A. Regev. 2004. A module map showing conditional activity of expression modules in cancer. *Nat. Genet.* 36: 1090–1098.
4. Fudenberg, G., G. Getz, M. Meyerson & L.A. Mirny. 2011. High order chromatin architecture shapes the landscape of chromosomal alterations in cancer. *Nat. Biotechnol.* 29: 1109–1113.
5. Hanahan, D. & R.A. Weinberg. 2011. Hallmarks of cancer: the next generation. *Cell* 144: 646–674.
6. Welch, J.S. *et al.* 2012. The origin and evolution of mutations in acute myeloid leukemia. *Cell* 150: 264–278.
7. Bric, A. *et al.* 2009. Functional identification of tumor-suppressor genes through an in vivo RNA interference screen in a mouse lymphoma model. *Cancer Cell* 16: 324–335.
8. Ngo, V.N. *et al.* 2006. A loss-of-function RNA interference screen for molecular targets in cancer. *Nature* 441: 106–110.
9. Mavrakis, K.J. *et al.* 2010. Genome-wide RNA-mediated interference screen identifies miR-19 targets in Notch-induced T-cell acute lymphoblastic leukaemia. *Nat. Cell Biol.* 12: 372–379.
10. Zuber, J. *et al.* 2011. RNAi screen identifies Brd4 as a therapeutic target in acute myeloid leukaemia. *Nature* 478: 524–528.
11. Zender, L. *et al.* 2008. An oncogenomics-based in vivo RNAi screen identifies tumor suppressors in liver cancer. *Cell* 135: 852–864.
12. Egle, A., A.W. Harris, M.L. Bath, *et al.* 2004. VavP-Bcl2 transgenic mice develop follicular lymphoma preceded by germinal center hyperplasia. *Blood* 103: 2276–2283.
13. Wendel, H.G. *et al.* 2004. Survival signalling by Akt and eIF4E in oncogenesis and cancer therapy. *Nature* 428: 332–337.
14. Schatz, J.H. *et al.* 2011. Targeting cap-dependent translation blocks converging survival signals by AKT and PIM kinases in lymphoma. *J. Exp. Med.* 208: 1799–1807.
15. Harris, A.W. *et al.* 1988. The E mu-myc transgenic mouse. A model for high-incidence spontaneous lymphoma and leukemia of early B cells. *J. Exp. Med.* 167: 353–371.
16. Cattoretti, G. *et al.* 2005. Deregulated BCL6 expression recapitulates the pathogenesis of human diffuse large B cell lymphomas in mice. *Cancer Cell* 7: 445–455.
17. Gladden, A.B., R. Woolery, P. Aggarwal, *et al.* 2006. Expression of constitutively nuclear cyclin D1 in murine lymphocytes induces B-cell lymphoma. *Oncogene* 25: 998–1007.
18. Tsujimoto, Y., J. Cossman, E. Jaffe & C.M. Croce. 1985. Involvement of the bcl-2 gene in human follicular lymphoma. *Science* 228: 1440–1443.
19. Kridel, R., L.H. Sehn & R.D. Gascoyne. 2012. Pathogenesis of follicular lymphoma. *J. Clin. Invest.* 122: 3424–3431.
20. Dave, S.S. *et al.* 2004. Prediction of survival in follicular lymphoma based on molecular features of tumor-infiltrating immune cells. *N. Engl. J. Med.* 351, 2159–2169.
21. Martinez-Climent, J.A. *et al.* 2003. Transformation of follicular lymphoma to diffuse large cell lymphoma is associated with a heterogeneous set of DNA copy number and gene expression alterations. *Blood* 101: 3109–3117.
22. Lo Coco, F. *et al.* 1993. p53 mutations are associated with histologic transformation of follicular lymphoma. *Blood* 82: 2289–2295.
23. Nanjangud, G. *et al.* 2007. Molecular cytogenetic analysis of follicular lymphoma (FL) provides detailed characterization of chromosomal instability associated with the t(14;18)(q32;q21) positive and negative subsets and histologic progression. *Cytogenet. Genome Res.* 118: 337–344.
24. Cheung, K.J. *et al.* 2009. Genome-wide profiling of follicular lymphoma by array comparative genomic hybridization reveals prognostically significant DNA copy number imbalances. *Blood* 113: 137–148.
25. Colombat, P. *et al.* 2001. Rituximab (anti-CD20 monoclonal antibody) as single first-line therapy for patients with follicular lymphoma with a low tumor burden: clinical and molecular evaluation. *Blood* 97: 101–106.
26. Maloney, D.G. 2012. Anti-CD20 antibody therapy for B-cell lymphomas. *N. Engl. J. Med.* 366: 2008–2016.
27. Solal-Celigny, P. 2006. Follicular lymphoma international prognostic index. *Curr. Treat. Options Oncol.* 7: 270–275.
28. Federico, M. *et al.* 2009. Follicular lymphoma international prognostic index 2: a new prognostic index for follicular lymphoma developed by the international follicular lymphoma prognostic factor project. *J. Clin. Oncol.* 27: 4555–4562.

29. Morin, R.D. *et al.* 2011. Frequent mutation of histone-modifying genes in non-Hodgkin lymphoma. *Nature* **476:** 298–303.

30. Yap, D.B. *et al.* 2011. Somatic mutations at EZH2 Y641 act dominantly through a mechanism of selectively altered PRC2 catalytic activity, to increase H3K27 trimethylation. *Blood* **117:** 2451–2459.

31. Challa-Malladi, M. *et al.* 2011. Combined genetic inactivation of beta2-Microglobulin and CD58 reveals frequent escape from immune recognition in diffuse large B cell lymphoma. *Cancer Cell* **20:** 728–740.

32. Davis, R.E. *et al.* 2010. Chronic active B-cell-receptor signalling in diffuse large B-cell lymphoma. *Nature* **463:** 88–92.

33. Yunis, J.J. *et al.* 1987. Multiple recurrent genomic defects in follicular lymphoma: a possible model for cancer. *N. Engl. J. Med.* **316:** 79–84.

34. Compagno, M. *et al.* 2009. Mutations of multiple genes cause deregulation of NF-kappaB in diffuse large B-cell lymphoma. *Nature* **459:** 717–721.

35. Kato, M. *et al.* 2009. Frequent inactivation of A20 in B-cell lymphomas. *Nature* **459:** 712–716.

36. Mandelbaum, J. *et al.* 2010. BLIMP1 is a tumor suppressor gene frequently disrupted in activated B cell-like diffuse large B cell lymphoma. *Cancer Cell* **18:** 568–579.

37. Dawson, D.W. *et al.* 2007. Global DNA methylation profiling reveals silencing of a secreted form of Epha7 in mouse and human germinal center B-cell lymphomas. *Oncogene* **26:** 4243–4252.

38. Pasquale, E.B. 2010. Eph receptors and ephrins in cancer: bidirectional signalling and beyond. *Nat. Rev. Cancer* **10:** 165–180.

39. Pasquale, E.B. 2008. Eph-ephrin bidirectional signaling in physiology and disease. *Cell* **133:** 38–52.

40. Depaepe, V. *et al.* 2005. Ephrin signalling controls brain size by regulating apoptosis of neural progenitors. *Nature* **435:** 1244–1250.

41. Rashid, T. *et al.* 2005. Opposing gradients of ephrin-As and EphA7 in the superior colliculus are essential for topographic mapping in the mammalian visual system. *Neuron* **47:** 57–69.

42. Holmberg, J., D.L. Clarke & J. Frisen. 2000. Regulation of repulsion versus adhesion by different splice forms of an Eph receptor. *Nature* **408:** 203–206.

43. Weinstein, I.B. 2002. Cancer. Addiction to oncogenes—the Achilles heal of cancer. *Science* **297:** 63–64.

44. Ventura, A. *et al.* 2007. Restoration of p53 function leads to tumour regression in vivo. *Nature* **445:** 661–665.

45. Luo, J., N.L. Solimini & S.J. Elledge. 2009. Principles of cancer therapy: oncogene and non-oncogene addiction. *Cell* **136:** 823–837.

46. Wang, J. *et al.* 2005. Downregulation of EphA7 by hypermethylation in colorectal cancer. *Oncogene* **24:** 5637–5647.

47. Kuang, S.Q. *et al.* 2010. Aberrant DNA methylation and epigenetic inactivation of Eph receptor tyrosine kinases and ephrin ligands in acute lymphoblastic leukemia. *Blood* **115:** 2412–2419.

48. Guan, M., C. Xu, F. Zhang & C. Ye. 2009. Aberrant methylation of EphA7 in human prostate cancer and its relation to clinicopathologic features. *Int. J. Cancer* **124:** 88–94.

49. Barretina, J. *et al.* 2010. Subtype-specific genomic alterations define new targets for soft-tissue sarcoma therapy. *Nat. Genet.* **42:** 715–721.

50. Peifer, M. *et al.* 2012. Integrative genome analyses identify key somatic driver mutations of small-cell lung cancer. *Nat. Genet.* **44:** 1104–1110.

51. Stransky, N. *et al.* 2011. The mutational landscape of head and neck squamous cell carcinoma. *Science* **333:** 1157–1160.

52. Hodis, E. *et al.* 2012. A landscape of driver mutations in melanoma. *Cell* **150:** 251–263.

53. Imielinski, M. *et al.* 2012. Mapping the hallmarks of lung adenocarcinoma with massively parallel sequencing. *Cell* **150:** 1107–1120.

54. Grasso, C.S. *et al.* 2012. The mutational landscape of lethal castration-resistant prostate cancer. *Nature* **487:** 239–243.

55. Rudin, C.M. *et al.* 2012. Comprehensive genomic analysis identifies SOX2 as a frequently amplified gene in small-cell lung cancer. *Nat. Genet.* **44:** 1111–1116.

56. Druker, B.J. *et al.* 2001. Efficacy and safety of a specific inhibitor of the BCR-ABL tyrosine kinase in chronic myeloid leukemia. *N. Engl. J. Med.* **344:** 1031–1037.

57. Poulikakos, P.I., C. Zhang, G. Bollag, et al. 2010. RAF inhibitors transactivate RAF dimers and ERK signalling in cells with wild-type BRAF. *Nature* **464:** 427–430.

58. Druker, B.J. et al. 2006. Five-year follow-up of patients receiving imatinib for chronic myeloid leukemia. *N. Engl. J. Med.* **355:** 2408–2417.

59. Al-Lazikani, B., U. Banerji & P. Workman. 2012. Combinatorial drug therapy for cancer in the post-genomic era. *Nat. Biotechnol.* **30:** 679–692.

60. Scuoppo, C. *et al.* 2012. A tumour suppressor network relying on the polyamine-hypusine axis. *Nature* **487:** 244–248.

61. Solimini, N.L. *et al.* 2012. Recurrent hemizygous deletions in cancers may optimize proliferative potential. *Science* **337:** 104–109.

62. Nijhawan, D. *et al.* 2012. Cancer vulnerabilities unveiled by genomic loss. *Cell* **150:** 842–854.

63. Ding, L., G. Getz, D.A. Wheeler, *et al.* 2008. Somatic mutations affect key pathways in lung adenocarcinoma. *Nature* **455:** 1069–1075.

64. Seshagiri, S., E.W. Stawiski, S. Durinck, *et al.* 2012. Recurrent R-spondin fusions in colon cancer. *Nature* **488:** 660–664.

65. Robinson, G., M. Parker, T.A. Kranenburg, *et al.* 2012. Novel mutations target distinct subgroups of medulloblastoma. *Nature* **488:** 43–48.

66. Stephens, P.J., P.S. Tarpey, H. Davies, *et al.* 2012. The landscape of cancer genes and mutational processes in breast cancer. *Nature* **486:** 400–404.

67. Guo, G., Y. Gui, S. Gao, *et al.* 2011. Frequent mutations of genes encoding ubiquitin-mediated proteolysis pathway components in clear cell renal cell carcinoma. *Nat. Genet.* **44:** 17–19.

68. Banerji, S., K. Cibulskis, C. Rangel-Escareno, *et al.* 2012. Sequence analysis of mutations and translocations across breast cancer subtypes. *Nature* **486:** 405–409.

69. Barbieri, C.E., S.C. Baca, M.S. Lawrence, *et al.* 2012. Exome sequencing identifies recurrent SPOP, FOXA1 and MED12 mutations in prostate cancer. *Nat. Genet.* **44:** 685–689.

70. Jones, D.T., N. Jäger, M. Kool, *et al.* 2012. Dissecting the genomic complexity underlying medulloblastoma. *Nature* **488:** 100–105.

Ann. N.Y. Acad. Sci. ISSN 0077-8923

ANNALS OF THE NEW YORK ACADEMY OF SCIENCES
Issue: *Blavatnik Awards for Young Scientists 2012*

RNA polymerase: in search of promoters

Andrey Feklistov

Laboratory of Molecular Biophysics, The Rockefeller University, New York, New York

Address for correspondence: Andrey Feklistov, The Rockefeller University, Laboratory of Molecular Biophysics, Box 224, 1230 York Avenue, New York, NY 10065. afeklistov@rockefeller.edu

Transcription initiation is a key event in the regulation of gene expression. RNA polymerase (RNAP), the central enzyme of transcription, is able to efficiently locate promoters in the genome, carry out promoter opening, and initiate RNA synthesis. All the substeps of transcription initiation are subject to complex cellular regulation. Understanding the molecular details of each step in the promoter-opening pathway is essential for a complete mechanistic and quantitative picture of gene expression. In this minireview, primarily using bacterial RNAP as an example, I briefly summarize some of the key recent advances in our understanding of the mechanisms of promoter search and promoter opening.

Keywords: RNA polymerase; transcription; promoter; protein–DNA recognition

Introduction

Transcription is the basis for decoding genetic information stored in DNA. Core RNA polymerase (subunit composition $\alpha_2\beta\beta'\omega$) is responsible for all cellular transcription in bacteria.[1] Specific transcription initiation at promoter sites requires an additional σ-subunit. Association of core and σ yields the holoenzyme capable of locating promoter sequences, opening DNA to form a transcription bubble, and initiating RNA synthesis.[2–4] The primary σ-factors (σ^{70} in *Escherichia coli*) feature four structural domains and direct core RNAP to the majority of promoters active during log-phase growth, whereas alternative σ-factors control specialized promoters activated in response to environmental and intracellular signals.[5]

RNAP holoenzyme binds DNA along an extensive, positively charged interface formed by various regions of all RNAP subunits (except ω) (Fig. 1A). Some parts of the interface are engaged in nonspecific DNA binding, primarily with the sugar – phosphate DNA backbone, whereas others ensure specific readout of the DNA sequence by interacting with the bases. The overall shape of RNAP resembles a crab claw: the two pincers form the active site cleft with the catalytic Mg^{2+} located deep in the cleft.[6] The backside of the holoenzyme where the

σ-subunit is bound features patches of positively charged surface carrying out promoter recruitment through nonspecific interactions as well as read-out of dsDNA (double stranded DNA) shape and sequence. After the initial positioning of RNAP on promoter DNA is achieved, a sharp kink (possible after nucleation of melting) brings the downstream DNA in contact with the active site cleft. The cleft is too narrow to accommodate dsDNA, so in order to reach the active site, DNA must unwind. The melting is achieved via interactions of DNA strands with the positively charged surface of the cleft, through both nonspecific contacts to the DNA backbone and specific interactions with individual bases in the ssDNA (single stranded DNA).

Analysis of promoter sequences revealed several conserved elements important for RNAP binding. The most common and most highly conserved are two hexamers centered 35 and 10 bp upstream of the transcription start site (+1): the −10 element (TATAAT) and the −35 element (TTGACA)[7]-recognized by domain 2 and 4 of σ, respectively.[8,9] Additional promoter elements include the extended −10 element (TG)[10] and the discriminator (GGGA),[11,12] also recognized by σ-subunit (domain 3 and 2, respectively).[13,14] Core subunits also provide DNA-binding specificity and are involved in recognition of the upstream promoter

doi: 10.1111/nyas.12197

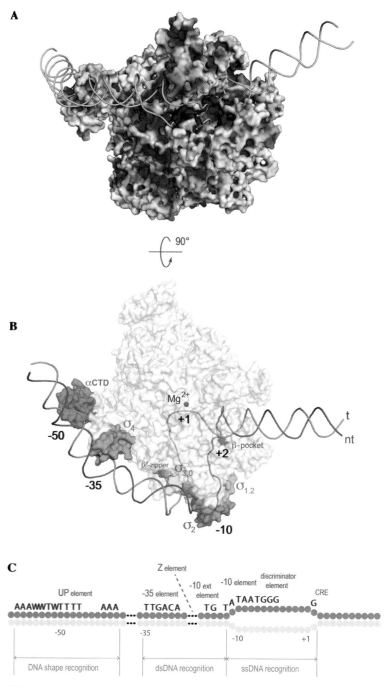

Figure 1. Model of the open promoter complex. (A) View showing electrostatic surface potential of RNA polymerase (RNAP) in the open promoter complex (red, negative; blue, positive charge). DNA (golden) is bound across a positively charged path on RNAP. RNAP engages the upstream region of promoter DNA (left) in sequence-specific recognition of dsDNA promoter elements. The melted part of the promoter bubble (right) is recognized through sequence-specific contacts with ssDNA. (B) View showing sequence-specific promoter elements and parts of RNAP recognizing them. RNAP shown as a gray transparent surface, except patches involved in sequence-specific recognition of promoter DNA. DNA backbone is outlined (nontemplate strand, blue; template strand, gray). Model was created by combining coordinates from PDB IDs 4G7H, 3UGO, 1LB2, and 1L9Z. (C) Promoter motifs recognized by RNAP holoenzyme with primary σ-factors. Blue circles represent nucleotides of the nontemplate DNA strand, light gray, template strand. W = A or T. The position with respect to the transcription start-site (+1) is denoted below.

(UP)-element, Z-element, and core recognition element (CRE)[14–16] (Fig. 1C).

With the exception of the indispensable –10 element, other promoter elements may or may not be present. Each element plays its role at a certain step(s) of the promoter-opening pathway as it is recognized by RNAP: some in dsDNA form as they recruit RNAP to the promoter region, while others in ssDNA form concurrently with melting. In fact, sequence-specific ssDNA recognition in the region undergoing melting initiates and drives promoter opening. The DNA-binding surfaces of RNAP, therefore, can be viewed as a collection of DNA-binding sites with different roles and DNA specificities that come into play in a concerted fashion. In the following, I will review recent advances in our understanding of how RNAP finds promoters in the vastness of the genome and how promoter melting occurs, leading to initiation of RNA synthesis.

Promoter search

DNA-binding proteins are able to locate their target sites in an overwhelming excess of non-target DNA. The impressive speed and efficiency of this search, given the size of the macromolecules involved, cannot be explained by simple diffusion in the cytoplasm.[17] This needle-in-the-haystack problem has puzzled molecular biologists for decades, and the mechanisms utilized by proteins in search of their target sites on genomic DNA are still being elucidated. To explain this paradox, Berg, Winter, and von Hippel suggested that DNA-binding proteins first use their nonspecific DNA-binding affinity to arrive at any binding site on the DNA and in the next step search for their target site by means of thermal diffusion along DNA (reduced available volume for this search explains fast target location).[18,19] Along a short DNA stretch, a protein can slide freely or translocate through a series of microscopic dissociation–reassociation events, whereas for larger DNA lengths the search could be manifested through intersegment transfer.

In the case of RNAP, the location of its binding site (promoter search) is perhaps the most enigmatic step in transcription initiation, due to the short-lived nature of the search intermediates. For the bacterial RNAP, the task of locating promoters within an average-sized genome translates into finding a fraction of sequences comprising just a small percentage of the available DNA in the cell.[20] Evidence for 1D sliding of RNAP along DNA has been obtained both in bulk biochemical assays[21,22] and in single-molecule experiments.[23] More specifically, RNAP has been shown to track a DNA groove as it moves along in search of a promoter site.[24] Time-resolved scanning force microscopy revealed that both 1D sliding and 3D hopping accompany RNAP movement along DNA,[25] whereas recent *in vivo* fluorescent microscopy studies suggest that both mechanisms may operate in living cells.[26] Despite the ample evidence that RNAP can slide along the DNA while searching for the promoter site, the observed promoter-association kinetic parameters are perfectly explained by a 3D-diffusion mechanism alone, and a recent study argued that 3D collisions may indeed be the only mechanism operating in the cell, given the high *in vivo* concentrations of RNAP holoenzyme.[27] It is still possible that low-copy transcription factors or RNAPs with alternative σ-factors (with just a few tens of molecules per cell) resort to facilitated diffusion in order to reach their target sites.

The structure of the bacterial nucleoid and the distribution of RNAP are dynamic and can be influenced by environmental conditions.[28] Similar to transcription factories in eukaryotes, bacteria can organize promoters in specific cellular locations, which in turn can modulate promoter strength.[29] Active promoters often reside within regions of the bacterial genome containing multiple overlapping promoter-like sequences, which could play roles in channeling RNAP into the promoter region.[30] For eukaryotic transcription factors, a mechanism was proposed that would rapidly engage a "treadmilling" transcription factor (i.e., in a state of continual binding and dissociation from its target site) and convert it to a more stable binding state, allowing for a clutch-like genomic response to developmental or environmental cues.[31] The latter work also introduced an important methodology to study the binding dynamics of a transcription factor as a true predictor of its strength. Studies of how the presence of additional promoter-like sequences affect RNAP binding turnover dynamics at true promoter sites could shed light on their role and explain the abundance of such sequences in bacterial genomes.

Base excision repair (BER) proteins, a classic model for the studies of target site search process, are able to locate isolated damaged bases in the genome with an efficiency that cannot be explained even by facilitated diffusion in a restricted volume. It was proposed that redox-active [4Fe–4S] clusters present in some DNA repair proteins are able to sense charge-conducting properties of DNA (electron propagation along an intact base stack that would be disrupted in the case of a DNA lesion) and dramatically improve the efficiency of damage detection via this mechanism.[32] In addition to DNA repair factors, Fe–S clusters were found to be essential components of various nucleic acid–processing enzymes such as DNA polymerases, helicases, glycosylases, primases, nucleases, and transcription factors.[33] RNAPs from some archea, plants, and protozoa contain an Fe–S cluster that is required for RNAP assembly and has been proposed to play a role in sensing redox state of the cell.[34] The location of the cluster near DNA in the modeled RNAP–promoter complex and its high conservation among several evolutionary distant RNAPs suggest the attractive but, at this point, speculative possibility that it may be involved in DNA-mediated redox signaling either directly or via a transcription factor.

The initial phase of promoter search by the bacterial RNAP may involve indirect readout of DNA (shape recognition). Variations in DNA sequence create unique conformational signatures with distinct geometrical helix parameters and deformability depending on local patterns of interactions between stacked bases. Whole-genome analyses argue that the topographical landscape of DNA molecular shape formed as a result of these interactions is conserved and can be subject to evolutionary selection[35] much like protein shape. This can provide an efficient means for fast shape readout.[36] For example, RNAP binding to UP-element involves recognition of a narrow minor groove,[15,37] whereas –10 hexamer DNA was proposed to have altered structure even in the absence of DNA-binding proteins.[38,39]

Although the quest for the characterization of the elusive promoter search intermediates is ongoing, one should keep in mind the highly mobile nature of the DNA helix, in which base flipping and noncanonical base pairing occur frequently and can be recognized by proteins. The detection of transient Hoogsteen base pairs forming within duplex DNA[40] points to the possibility for multiple layers of a protein–DNA readout code in addition to simple linear sequence.

In many cases, the promoter search by RNAP is modulated by the actions of activators and repressors (reviewed in Ref. 41). These protein factors can modify the RNAP DNA-binding surface by either supplying it with additional DNA-recognition determinants or by blocking existing surfaces. Alternatively, the modulating factors can change the conformation of the promoter DNA by either making it more attractive to RNAP or by obstructing existing promoter sites. The majority of activators and repressors act early in the initiation pathway (usually at the promoter recruitment phase), although recent studies of the bacteriophage T7 Gp2 inhibitor and the *Mycobacterium tuberculosis* transcriptional modulator CarD argue that these factors affect the promoter-opening step.[42,43]

Promoter opening

Following the recruitment phase of the promoter search, involving recognition of DNA shape and dsDNA sequence, RNAP unwinds about 1.3 turns of the DNA (from –11 to +3), forming the open promoter complex. At this stage, RNAP specifically recognizes individual bases in the nontemplate strand of the promoter DNA—this activity underlies the melting capabilities of RNAP because the contacts with ssDNA bases can only be established during or after DNA unwinding.

Promoter melting is triggered by the recognition of the –10 element, which occurs as the two most conserved bases of the element (A_{-11} and T_{-7}) are flipped out of the dsDNA and into complementary protein pockets of the $\sigma 2$ domain.[9,14,44] Structural modeling and biochemical data suggest a hypothetical timeline of this key event in promoter opening (Fig. 2): directed by upstream sequence-specific promoter–RNAP contacts and electrostatic interactions with the sugar–phosphate backbone, the –10 element DNA is loaded in a shallow positively charged trough formed by the surface of $\sigma 2$, $\sigma 3$, and parts of the β subunit. A prominent Trp residue (σ W433 in *E. coli*) located at the bottom of the trough precludes binding of an intact B-form DNA helix, acting as a wedge disrupting the –11 base pair and initiating the flipping of –11A into its pocket.[9,45] This absolutely conserved Trp may be acting, therefore, as a functional analog of the "interrogating residue" of DNA-binding proteins that recognize flipped-out bases.[46,47]

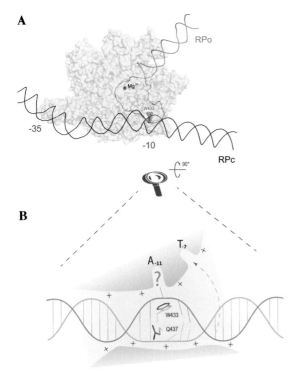

Figure 2. Structural modeling of –10 element recognition as the first step in promoter opening. (A) Schematic comparison of closed (RPc) and open (RPo) promoter complexes and suggested wedge role for W433 as an initiator of promoter melting. RNAP shown as a gray semitransparent surface. DNA in RPc is shown in black, downstream portion of promoter DNA after the melting (RPo) is green. The wedge W433 is highlighted in red. Catalytic Mg^{2+}, purple sphere; –35 and –10 promoter elements are labeled. (B) Schematic close-up of the first step in –10 element recognition. DNA directed in the shallow trough on RNAP surface (green) via a steric and electrostatic fit. Nontemplate strand is shown in blue, template strand is gray. The helix invasion by W433 disrupts one of the base pairs and flips out the nontemplate base into A_{-11} pocket, thereby making the upstream neighboring template-strand base accessible for H-bonding with Q437 (this amino acid residue was implicated in –12 base pair recognition in genetic screens).[48,49] Following the recognition of the $T_{-12}A_{-11}$ step, DNA untwisting will proceed downstream, accompanied by T_{-7} flipping out into the respective σ-pocket (shown by red dashed arrow).

Remarkably, even though dsDNA recognition elements recruit RNAP to the promoter and align the –10 element with the recognition surface of σ2, specific recognition of the –10 element and initiation of melting (at least *in vitro*) has been shown to occur in their absence. Promoter fragments having only the –10 and discriminator elements (both recognized in ssDNA form) support transcription *in vitro*,[11] and DNA fragments with only the –10 element are recog-

nized by RNAP through specific interaction with the nontemplate strand.[9,50] These examples imply that RNAP distorts the DNA helix as it searches for the –10 element within dsDNA. The prominent position of Trp433 at the cusp of the protein–DNA interface at the origin of the melting suggests that as the dsDNA is threaded through the trough during the search process, the Trp wedge can flip bases out to be sampled in the A_{-11} pocket (Fig. 2).

Using structural modeling along with biochemical evidence, Feklistov and Darst[9] challenged the previous concept of sequence-specific recognition of the double-stranded –10 element in the closed promoter complex, suggesting that binding of RNAP to the –10 element DNA involves an intermediate with locally distorted duplex where the DNA helix is preopened to facilitate readout of the flipped bases. The early intermediates of promoter opening observed on various promoters have been termed closed on the basis of their nonreactivity toward MnO_4^- treatment (an assay used to reveal unstacked, solvent-exposed thymine bases). This technique fails to detect opening in cases where the thymines in the melted region remain stacked and/or protected by contacts with the protein. Therefore, new methods for assessing the state of the DNA helix in early melting intermediates need to be developed.

After initiation of melting at the A_{-11} position, the promoter bubble grows in the downstream direction. Sequence-specific recognition of additional ssDNA elements (discriminator, –6 to –4; and CRE, –4 to +2) at some promoters may facilitate promoter opening at this stage (Fig. 1C). Therefore, the checkpoints in the process of promoter search do not stop at the recruitment phase but continue all the way to the formation of the fully open promoter bubble. Even when the bubble is fully formed, sequence-dependent conformational fluctuations of the open complex can fine-tune the choice of the transcription start site.[51]

The exact sequence of events leading to the formation of the open promoter bubble is currently a subject of debate. Data obtained at the bacteriophage λ PR promoter at subphysiological temperatures (used to slow down the opening reaction) argue that after establishing upstream promoter contacts, RNAP readily bends promoter DNA (at about 90°) at the –10 element and inserts the downstream duplex in the active site cleft.[52] In the next slow step, RNAP opens the DNA and readjusts contacts with emerging stretches of ssDNA as well as with the downstream DNA duplex.[53–55] Real time X-ray–generated hydroxyl-radical footprinting data at the bacteriophage T7A1 promoter under physiological temperatures suggest that DNA opening originates outside of the cleft whereas DNA bends and enters the cleft later in the pathway.[56,57] It is not impossible that even for a particular promoter both pathways may coexist. It is possible that, depending on promoter sequence, one or the other pathway may prevail, which could present an opportunity for differential regulation *in vivo*. Notably, even for the same promoter sequence, DNA opening pathway can go through different structural intermediates depending on the experimental temperature.[57]

Overall, promoter melting is driven by RNAP affinity toward the final state (i.e., the conformation of promoter DNA existing in RPo). At the opening step, RNAP can be envisioned as an isomerization machine utilizing binding free energy to bend promoter DNA around its surface and unwind about 13 base pairs of the dsDNA, placing the template strand near the active site ready for coding of the transcript sequence.[52] Opening of the promoter DNA is accompanied by closing of the RNAP clamp,[58] allowing RNAP to acquire a tight grip on the downstream DNA to assure processivity of transcription. This process may involve refolding of several structural domains of RNAP at the downstream parts of the cleft.[52,55] The driving force of this dramatic rearrangement is supplied by electrostatic and steric complementarity between the positively charged surface of RNAP and the negatively charged DNA backbone, whereas the accuracy ensuring precise register of transcription comes from sequence-specific contacts to the dsDNA and ssDNA promoter elements (Fig. 1B).

Conclusions

Between the moment of association of the RNAP core enzyme with one of the cellular σ-factors and the moment of synthesis of the first phosphodiester bond, RNAP is engaged in a complex multistep process of promoter search. Each substep of this process can be modulated by regulatory protein factors or small molecules.[46,59] The transcriptional output of a promoter is determined not just by how efficiently it is located and melted by RNAP, but also by the ease of promoter escape. Mechanistic understanding of the

role of individual promoter elements in determining promoter output is a prerequisite for building a detailed quantitative model of bacterial gene expression. Recent structural and biochemical studies have advanced our understanding of the open promoter complex organization, although many important promoter search/opening intermediates still await characterization.

Acknowledgments

I would like to thank Ruth Saecker and Anastasia Sevostyanova for helpful comments on this manuscript. I am especially grateful to Seth A. Darst for inspiring mentorship, engaging discussions and support. My postdoctoral research was supported, in part, by a Merck Postdoctoral Fellowship at the Rockefeller University and NIH R01 GM053759 Grant to Seth A. Darst.

Conflicts of interest

The author declares no conflicts of interest.

References

1. Darst, S.A. 2001. Bacterial RNA polymerase. *Curr. Opin. Struct. Biol.* **11:** 155–162.
2. Vassylyev, D.G., S.-I. Sekine, O. Laptenko, *et al.* 2002. Crystal structure of a bacterial RNA polymerase holoenzyme at 2.6 A resolution. *Nature* **417:** 712–719.
3. Murakami, K.S., S. Masuda & S.A. Darst. 2002. Structural basis of transcription initiation: RNA polymerase holoenzyme at 4 A resolution. *Science* **296:** 1280–1284.
4. Murakami, K.S. 2013. The X-ray crystal structure of *Escherichia Coli* RNA polymerase sigma70 holoenzyme. *J. Biol. Chem.* **288:** 9126–9134.
5. Gruber, T.M. & C.A. Gross. 2003. Multiple sigma subunits and the partitioning of bacterial transcription space. *Annu. Rev. Microbiol.* **57:** 441–466.
6. Murakami, K.S., & S.A. Darst. 2003. Bacterial RNA polymerases: the wholo story. *Curr. Opin. Struct. Biol.* **13:** 31–39.
7. Shultzaberger, R.K., Z. Chen, K.A. Lewis & T.D. Schneider. 2007. Anatomy of *Escherichia coli* sigma70 promoters. *Nucleic Acids Res.* **35:** 771–788.
8. Campbell, E.A., O. Muzzin, M. Chlenov, *et al.* 2002. Structure of the bacterial RNA polymerase promoter specificity sigma subunit. *Mol. Cell* **9:** 527–539.
9. Feklistov, A. & S.A. Darst. 2011. Structural basis for promoter –10 element recognition by the bacterial RNA polymerase σ subunit. *Cell* **147:** 1257–1269.
10. Barne, K.A., J.A. Bown, S.J.W. Busby & S.D. Minchin. 1997. Region 2.5 of the *Escherichia coli* RNA polymerase sigma70 subunit is responsible for the recognition of the "extended-10" motif at promoters. *EMBO J.* **16:** 4034–4040.
11. Feklistov, A., N. Barinova, A. Sevostyanova, *et al.* 2006. A basal promoter element recognized by free RNA polymerase

σ subunit determines promoter recognition by RNA polymerase holoenzyme. *Mol. Cell* **23**: 97–107.

12. Haugen, S.P., M.B. Berkmen, W. Ross, *et al.* 2006. rRNA promoter regulation by nonoptimal binding of sigma region 1.2: an additional recognition element for RNA polymerase. *Cell* **125**: 1069–1082.

13. Murakami, K.S., S. Masuda, E.A. Campbell, *et al.* 2002b. Structural basis of transcription initiation: an RNA polymerase holoenzyme-DNA complex. *Science* **296**: 1285–1290.

14. Zhang, Y., Y. Feng, S. Chatterjee, *et al.* 2012. Structural basis of transcription initiation. *Science* **338**: 1076–1080.

15. Benoff, B., H. Yang, C.L. Lawson, *et al.* 2002. Structural basis of transcription activation: the CAP-alpha CTD–DNA complex. *Science* **297**: 1562–1566.

16. Yuzenkova, Y., V.R. Tadigotla, K. Severinov & N. Zenkin. 2011. A new basal promoter element recognized by RNA polymerase core enzyme. *EMBO J.* **30**: 3766–3775.

17. Riggs, A.D., S. Bourgeois & M. Cohn. 1970. The lac repressor–operator interaction. 3. Kinetic studies. *J. Mol. Biol.* **53**: 401–417.

18. Berg, O.G., R.B. Winter & P.H. von Hippel. 1981. Diffusion-driven mechanisms of protein translocation on nucleic acids. 1. Models and theory. *Biochemistry* **20**: 6929–6948.

19. Hippel von, P.H. & O.G. Berg. 1989. Facilitated target location in biological systems. *J. Biol. Chem.* **264**: 675–678.

20. Wang, F. & E.C. Greene. 2011. Single-molecule studies of transcription: from one RNA polymerase at a time to the gene expression profile of a cell. *J. Mol. Biol.* **412**: 814–831.

21. Park, C.S., F.Y. Wu & C.W. Wu. 1982. Molecular mechanism of promoter selection in gene transcription. II. Kinetic evidence for promoter search by a one-dimensional diffusion of RNA polymerase molecule along the DNA template. *J. Biol. Chem.* **257**: 6950–6956.

22. Ricchetti, M., W. Metzger & H. Heumann. 1988. One-dimensional diffusion of *Escherichia coli* DNA-dependent RNA polymerase: a mechanism to facilitate promoter location. *Proc. Natl. Acad. Sci. USA* **85**: 4610–4614.

23. Kabata, H., O. Kurosawa, I. Arai, *et al.* 1993. Visualization of single molecules of RNA polymerase sliding along DNA. *Science* **262**: 1561–1563.

24. Sakata-Sogawa, K. & N. Shimamoto. 2004. RNA polymerase can track a DNA groove during promoter search. *Proc. Natl. Acad. Sci. USA* **101**: 14731–14735.

25. Bustamante, C., M. Guthold, X. Zhu & G. Yang. 1999. Facilitated target location on DNA by individual *Escherichia coli* RNA polymerase molecules observed with the scanning force microscope operating in liquid. *J. Biol. Chem.* **274**: 16665–16668.

26. Bratton, B.P., R.A. Mooney & J.C. Weisshaar. 2011. Spatial distribution and diffusive motion of RNA polymerase in live *Escherichia coli*. *J. Bacteriol.* **193**: 5138–5146.

27. Wang, F., S. Redding, I.J. Finkelstein, *et al.* 2012. The promoter-search mechanism of *Escherichia coli* RNA polymerase is dominated by three-dimensional diffusion. *Nat. Struct. Mol. Biol.* **20**: 174–181.

28. Cabrera, J.E. & D.J. Jin. 2003. The distribution of RNA polymerase in *Escherichia coli* is dynamic and sensitive to environmental cues. *Mol. Microbiol.* **50**: 1493–1505.

29. Sánchez-Romero, M.-A., D.J. Lee, E. Sánchez-Morán & S.J.W. Busby. 2012. Location and dynamics of an active promoter in *Escherichia coli* K-12. *Biochem. J.* **441**: 481–485.

30. Huerta, A.M. & J. Collado-Vides. 2003. Sigma70 promoters in *Escherichia coli*: specific transcription in dense regions of overlapping promoter-like signals. *J. Mol. Biol.* **333**: 261–278.

31. Lickwar, C.R., F. Mueller, S.E. Hanlon, *et al.* 2012. Genome-wide protein-DNA binding dynamics suggest a molecular clutch for transcription factor function. *Nature* **484**: 251–255.

32. Sontz, P.A., N.B. Muren & J.K. Barton. 2012. DNA charge transport for sensing and signaling. *Acc. Chem. Res.* **45**: 1792–1800.

33. White, M.F. & M.S. Dillingham. 2012. Iron–sulphur clusters in nucleic acid processing enzymes. *Curr. Opin. Struct. Biol.* **22**: 94–100.

34. Hirata, A. & K.S. Murakami. 2009. Archaeal RNA polymerase. *Curr. Opin. Struct. Biol.* **19**: 724–731.

35. Parker, S.C.J., L. Hansen, H.O. Abaan, *et al.* 2009. Local DNA topography correlates with functional noncoding regions of the human genome. *Science* **324**: 389–392.

36. Rohs, R., X. Jin, S.M. West, *et al.* 2010. Origins of specificity in protein–DNA recognition. *Annu. Rev. Biochem.* **79**: 233–269.

37. Gourse, R.L., W. Ross & T. Gaal. 2000. UPs and downs in bacterial transcription initiation: the role of the alpha subunit of RNA polymerase in promoter recognition. *Mol. Microbiol.* **37**: 687–695.

38. Drew, H.R., J.R. Weeks & A.A. Travers. 1985. Negative supercoiling induces spontaneous unwinding of a bacterial promoter. *EMBO J.* **4**: 1025–1032.

39. Spassky, A., S. Rimsky, H. Buc & S.J.W. Busby. 1988. Correlation between the conformation of *Escherichia coli* –10 hexamer sequences and promoter strength: use of orthophenanthroline cuprous complex as a structural index. *EMBO J.* **7**: 1871–1879.

40. Nikolova, E.N., E. Kim, A.A. Wise, *et al.* 2011. Transient Hoogsteen base pairs in canonical duplex DNA. *Nature* **470**: 498–502.

41. Lee, D.J., S.D. Minchin & S.J.W. Busby. 2012. Activating transcription in bacteria. *Annu. Rev. Microbiol.* **66**: 125–152.

42. James, E., M. Liu, C. Sheppard, *et al.* 2012. Structural and mechanistic basis for the inhibition of *Escherichia coli* RNA polymerase by T7 Gp2. *Mol. Cell* **47**: 755–766.

43. Bae, B., E. Campbell, S. Darst, *et al.*, manuscripts in preparation.

44. Liu, X., D.A. Bushnell & R.D. Kornberg. 2011. Lock and key to transcription: σ–DNA interaction. *Cell* **147**: 1218–1219.

45. Tomsic, M., L. Tsujikawa, G. Panaghie, *et al.* 2001. Different roles for basic and aromatic amino acids in conserved region 2 of *Escherichia coli* sigma(70) in the nucleation and maintenance of the single-stranded DNA bubble in open RNA polymerase-promoter complexes. *J. Biol. Chem.* **276**: 31891–31896.

46. Lee, S., B.R. Bowman, Y. Ueno, *et al.* 2008. Synthesis and structure of duplex DNA containing the genotoxic nucleobase lesion N7-methylguanine. *J. Am. Chem. Soc.* **130**: 11570–11571.

47. Yi, C., B. Chen, B. Qi, *et al.* 2012. Duplex interrogation by a direct DNA repair protein in search of base damage. *Nat. Struct. Mol. Biol.* **19:** 671–676.

48. Kenney, T.J., K. York, P. Youngman & C.P.J. Jr. Moran. 1989. Genetic evidence that RNA polymerase associated with s^A factor uses a sporulation-specific promoter in *Bacillus subtilis. Proc. Natl. Acad. Sci. USA* **86:** 9109–9113.

49. Waldburger, C., T. Gardella, R. Wong & M.M. Susskind. 1990. Changes in conserved region 2 of *Escherichia coli* sigma 70 affecting promoter recognition. *J. Mol. Biol.* **215:** 267–276.

50. Niedziela-Majka, A. & T. Heyduk. 2005. Escherichia coli RNA polymerase contacts outside the -10 promoter element are not essential for promoter melting. *J. Biol. Chem.* **280:** 38219–38227.

51. Robb, N.C., T. Cordes, L.C. Hwang, *et al.* 2013. The transcription bubble of the RNA polymerase–promoter open complex exhibits conformational heterogeneity and millisecond-scale dynamics: implications for transcription start-site selection. *J. Mol. Biol.* **425:** 875–885.

52. Saecker, R.M., M.T. Record & P.L. deHaseth. 2011. Mechanism of bacterial transcription initiation: RNA polymerase-promoter binding, isomerization to initiation-competent open complexes, and initiation of RNA synthesis. *J. Mol. Biol.* **412:** 754–771.

53. Davis, C.A., C.A. Bingman, R. Landick, *et al.* 2007. Real-time footprinting of DNA in the first kinetically significant intermediate in open complex formation by *Escherichia coli* RNA polymerase. *Proc. Natl. Acad. Sci. USA* **104:** 7833–7838.

54. Gries, T.J., W.S. Kontur, M.W. Capp, *et al.* 2010. One-step DNA melting in the RNA polymerase cleft opens the initiation bubble to form an unstable open complex. *Proc. Natl. Acad. Sci.* **107:** 10418–10423.

55. Drennan, A., M. Kraemer, M. Capp, *et al.* 2012. Key roles of the downstream mobile jaw of *Escherichia coli* RNA polymerase in transcription initiation. *Biochemistry* **51:** 9447–9459.

56. Sclavi, B., E. Zaychikov, A. Rogozina, *et al.* 2005. Real-time characterization of intermediates in the pathway to open complex formation by *Escherichia coli* RNA polymerase at the T7A1 promoter. *Proc. Natl. Acad. Sci. USA* **102:** 4706–4711.

57. Rogozina, A., E. Zaychikov, M. Buckle, *et al.* 2009. DNA melting by RNA polymerase at the T7A1 promoter precedes the rate-limiting step at 37 degrees C and results in the accumulation of an off-pathway intermediate. *Nucleic Acids Res* **37:** 5390–5404.

58. Chakraborty, A., D. Wang, Y.W. Ebright, *et al.* 2012. Opening and closing of the bacterial RNA polymerase clamp. *Science* **337:** 591–595.

59. Haugen, S.P., W. Ross & R.L. Gourse. 2008. Advances in bacterial promoter recognition and its control by factors that do not bind DNA. *Nat. Rev. Microbiol.* **6:** 507–519.

Ann. N.Y. Acad. Sci. ISSN 0077-8923

Lessons about terminal differentiation from the specification of color-detecting photoreceptors in the *Drosophila* retina

Robert J. Johnston Jr.

Department of Biology, Johns Hopkins University, Baltimore, Maryland

Address for correspondence: Robert J. Johnston Jr., Department of Biology, Johns Hopkins University, 206 Mudd Hall, 3400 N. Charles Street, Baltimore, MD 21218-2685. robertjohnston@jhu.edu

Metazoans require highly diverse collections of cell types to sense, interpret, and react to the environment. Developmental programs incorporate deterministic and stochastic strategies in different contexts or different combinations to establish this multitude of cell fates. Precise genetic dissection of the processes controlling terminal photoreceptor differentiation in the *Drosophila* retina has revealed complex regulatory mechanisms required to generate differences in gene expression and cell fate. In this review, I discuss how a gene regulatory network interprets stochastic and regional inputs to determine the specification of color-detecting photoreceptor subtypes in the *Drosophila* retina. These combinatorial gene regulatory mechanisms will likely be broadly applicable to nervous system development and cell fate specification in general.

Keywords: stochastic; regulatory network; retina; feedforward loop; Rhodopsin

Introduction

How is the incredible diversity of cell types needed by organisms to function properly generated? This question has been at the core of much scientific research since the discovery of the first cell by Robert Hooke in the 1600s.[1] Though we have learned a great deal about signaling pathways and lineage-based mechanisms of specification, we are only beginning to understand how complex gene regulatory networks interpret stochastic inputs to diversify cell fates during development. Simple model organisms have provided insights into these fundamental processes and continue to reveal new molecular mechanisms of terminal differentiation. Though studied for over 100 years, the fly retina remains a powerful paradigm to identify and characterize new mechanisms of gene regulation. Here, I review the key mechanisms determining subtype specification of the color-detecting photoreceptors of the fly eye, compare their roles in other biological contexts, and discuss the use of stochastic gene expression as a complement to lineage- and signal-based inputs into cell fate differentiation.

The *Drosophila* retina contains approximately 800 ommatidia, or unit eyes. Each ommatidium contains eight photoreceptor cells (PRs), named R1–R8.[2] The outer PRs, R1–R6, express the broad wavelength–detecting Rhodopsin1 (Rh1) and are involved in motion detection (Fig. 1A).[3–5] Differential expression of color-detecting Rhodopsins (Rhs) in the inner PRs, R7 and R8, defines several different subtypes of ommatidia (Fig. 1A). The two main subtypes are called pale (**p**) and yellow (**y**), and are randomly distributed throughout the retina but occur in a **p:y** ratio of approximately 1:2 (Fig. 1B).[6–14] The **p** subtype contains **p**R7s that express UV-detecting Rh3 and **p**R8s that express blue-detecting Rh5 (Fig. 1C). The **y** subtype contains **y**R7s that express UV-detecting Rh4 and **y**R8s that express green-detecting Rh6 (Fig. 1D).

In addition to these two main, stochastically distributed ommatidial subtypes, the retina contains two additional subtypes that are regionally determined and appear to have specialized functions. In the dorsal third of the retina, Rh3 is coexpressed in Rh4-expressing **y**R7s, presumably sensitizing these

doi: 10.1111/nyas.12178

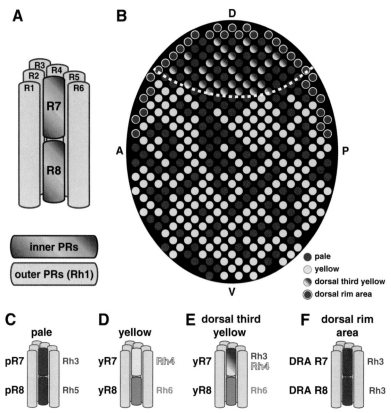

Figure 1. The four types and spatial distributions of ommatidial subtypes in the *Drosophila* retina. (A) Each ommatidium contains eight PRs including six motion-detecting outer PRs that express Rh1 (R1–R6 in gray) and two color-sensing inner PRs that express Rh3, Rh4, Rh5, or Rh6 (R7 and R8 in rainbow). (B) Shown are the distribution of ommatidial subtypes in the fly eye. Pale (purple circles) and yellow (yellow circles) subtypes are randomly distributed throughout the majority of the retina. In the dorsal third, pale and specialized dorsal third yellow (purple/yellow circles) subtypes are observed. At the dorsal rim of the retina, termed the dorsal rim area (DRA), DRA (purple circles with white outline) subtypes are found. (C) In the pale ommatidial subtype, Rh3 is expressed in **p**R7s, and Rh5 is expressed in **p**R8s. (D) In the yellow ommatidial subtype, Rh4 is expressed in **y**R7s, and Rh6 is expressed in **y**R8s. (E) In the dorsal third yellow ommatidial subtype, Rh3 and Rh4 are coexpressed in **y**R7s, and Rh6 is expressed in **y**R8s. (F) In the DRA subtype, Rh3 is expressed in both DRA R7s and DRA R8s.

PRs to detect the orientation of the sun (Fig. 1E).[15] At the dorsal rim of the retina (dorsal rim area, DRA), ommatidia are specialized for the detection of polarized light and express Rh3 in both R7s and R8s (Fig. 1F).[16,17]

Stochastic and regional inputs are interpreted by a complex gene regulatory network to determine the mosaic of ommatidial subtypes. The mechanisms controlling most color-detecting Rhs involve general activation in all PRs throughout the retina combined with repression to restrict expression to specific PR subtypes, and additional activation to generate robustness. I refer to robust expression as expression of a given Rh in every cell of a specific PR subtype. In this review, I describe the network

logic controlling Rh expression in the fly visual system, starting with the general activation mechanism and stepping through additional layers of regulation that limit expression to distinct cell types. I conclude with a comparison of the uses of these regulatory mechanisms in the fly eye and other systems and a discussion of the roles of stochastic cell fate specification phenomena during development.

General activation in all PRs and restriction to inner PRs

A pair of homeodomain proteins, Orthodenticle (Otd) and Pph13, activate expression of Rhs in all PRs of the eye, including all R7s and R8s and even motion-detecting outer PRs. In particular, Otd

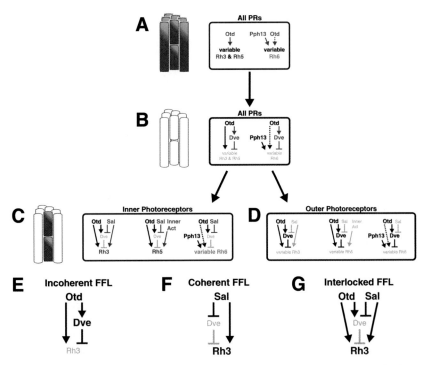

Figure 2. An incoherent FFL restricts expression of Rhs to inner PRs. (A) Otd activates Rh3 and Rh5, and Pph13 activates Rh6, in all PRs. This activation is nonrobust leading to variable levels of Rh protein in PRs. (B) Otd activates Dve that represses Rh3, Rh5, and Rh6 in all PRs. The interaction of these genes is termed an incoherent feedforward loop in which the upstream regulator Otd activates *rh* target genes but also activates the Dve repressor of the same target genes. (C) Sal represses Dve to allow for expression and induces robust expression of Rh3 in inner PRs. The interaction of Sal, Dve, and Rh3 is termed a coherent feedforward loop in which the upstream regulator Sal activates the *rh3* target gene and represses the Dve repressor of *rh3* to allow for expression. An unknown inner PR-specific factor replaces Sal to induce Rh5. (D) In the absence of Sal, the incoherent feedforward loop prevents aberrant expression of color-sensing Rhs in the motion-detecting outer PRs. (E) Incoherent FFL: Otd activates Rh3 but also activates Dve (a repressor of Rh3) yielding repression of Rh3. (F) Coherent FFL: Sal activates Rh3 and also represses Dve (a repressor of Rh3) allowing Rh3 expression. (G) Interlocked FFL: the incoherent FFL and coherent FFL share Dve as a critical node controlling Rh expression.

activates Rh3 and Rh5 whereas Pph13 activates Rh6 (Fig. 2A). Though this activating input occurs in all PRs, it is not potent, yielding variable, nonrobust expression levels of Rh protein (Fig. 2A).[18–21]

Otd triggers expression of the defective proventriculus (Dve) homeodomain protein that represses expression of Rh3, Rh5, and Rh6. Together, these genes form an incoherent feedforward loop (FFL) where the upstream regulator Otd directly activates target *rh* genes but also activates Dve, a direct repressor of the same *rh* target genes (Fig. 2B and E). Together, the incoherent FFL represses Rh3, Rh5, and Rh6 expression in all cells of the retina.[18] Incoherent FFLs have been shown to generate pulses of expression, accelerate responses to upstream inducers, control expression in a dose-dependent biphasic manner, and detect fold-changes in input levels.[22–26]

In contrast, this mechanism appears to provide a fail-safe against aberrant expression of color-detecting Rhs in motion-detecting outer PRs. It also sets up conditions where release of repression by Dve will allow expression of these Rhs.

Dve repression is relieved in inner PRs through the activity of the Spalt zinc finger transcription factors (Spalt major/Salm and Spalt-related/Salr, collectively referred to as Sal). Sal represses Dve to allow for expression of Rh3, Rh5, and Rh6 in R7s and R8s (Fig. 2C). Though this release of repression allows for expression, general activation by Otd and Pph13 only induces variable expression. Sal has a second function with Otd to activate robust Rh3 expression (i.e., expression in all inner PRs). The regulatory interactions among Sal, Dve, and Rh3 form a coherent FFL where Sal activates Rh3 and also represses

Dve, a repressor of Rh3 expression (Fig. 2C and F). For Rh5, an unknown inner PR-specific factor substitutes for Sal to activate Rh5 in all inner PRs (Fig. 2C). For Rh6, the R8-specific senseless (Sens) zinc finger protein takes the place of Sal to induce robust Rh6 expression in all R8s (Fig. 2C).[18,27,28] Coherent FFLs have been implicated in delaying responses and prolonging expression.[22,26,29] In the eye, the coherent FFL prevents repression by Dve and induces robust *rh* gene expression.

The incoherent FFL of Otd/Dve/Rh3 and the coherent FFL of Sal/Dve/Rh3 function together to form an interlocked FFL motif with Dve as a shared, critical node (Fig. 2C and G). This interlocked FFL acts as a binary switch for Rh expression. In the outer PRs, the incoherent FFL yields repression of the inner PR-specific Rhs (Fig. 2D). In the inner PRs, the coherent FFL counters the incoherent FFL by repressing Dve expression and inducing robust expression of Rh3, Rh5, and Rh6 (Fig. 2C).[18]

Restriction of Rh3 to R7s, and Rh5 and Rh6 to R8s

The interlocked FFL motif ensures that Rh3, Rh5, and Rh6 are expressed in inner PRs and repressed in outer PRs. The next step in the regulatory logic involves the restriction of Rh3 to R7s, and Rh5 and Rh6 to R8s. Sens directly represses Rh3 in R8s, restricting its expression to R7s (Fig. 3). The homeodomain protein Prospero (Pros) directly represses Rh5 and Rh6 in R7s, restricting their expression to R8 (Fig. 3).[18,30] Thus, Sens and Pros act as simple binary switches to restrict Rh expression to their proper PR type.

Regulation of rhodopsin expression in p and y subtypes

Up to this point, the regulatory logic controlling Rh3, Rh5, and Rh6 has been highly similar, requiring general activation in all PRs (Fig. 2A), repression in outer PRs by incoherent FFLs (Fig. 2D), derepression and robust activation in inner PRs by coherent FFLs (Fig. 2C), and restriction to inner PR types by binary switches (Fig. 3). Additionally, the regulatory mechanisms discussed thus far act uniformly for each given PR subtype in all ommatidia across the retina. For example, the incoherent FFL represses Rhs in all outer PRs throughout the retina. At the level of subtype specification (i.e., **p** versus **y** fate), the mechanism restricting expression of Rh3

to the **p**R7 subtype is dramatically different from the mechanism dictating expression of Rh5 and Rh6 in specific R8 subtypes. Furthermore, the **p** and **y** subtypes are randomly distributed throughout the retina, requiring a stochastic input to determine the binary fate choice.[8]

The stochastic input determining **p** versus **y** fate occurs in R7s and is then communicated to R8s in the same ommatidium. The key is the stochastic expression of the Spineless (Ss) PAS-bHLH transcription factor in R7s. In a random subset of R7s, Ss is expressed and functions with its ubiquitously expressed heterodimerization partner, Tango (Tgo), to reactivate Dve to repress Rh3 in **y**R7s despite the presence of the Otd and Sal activators (Fig. 4A and C). Ss also appears to directly activate expression of Rh4 (Fig. 4C). This simple activation of Rh4 by a single cell-specific transcription factor is unique among the inner PR-specific, color-detecting Rhs, which otherwise rely on complex mechanisms coupling general activation with cell-specific repression. In the subset of R7s lacking Ss, Otd and Sal induce the default **p**R7 fate including expression of Rh3 (Fig. 4A and B).[18,31,32]

In R7s, Ss also regulates the signal that is transduced to control the expression of Rh5 and Rh6 in subtypes in R8s. In **y**R7s, Ss/Tgo represses the signal, allowing for default **y**R8 fate including expression of Rh6 (Fig. 4C and F). In **p**R7s that lack Ss, the signal induces **p**R8 fate, including expression of Rh5 (Fig. 4B and E). In *ss* mutants, most R8s assume **p**R8 fate (Rh5 expression) whereas ectopic Ss expression allows for **y**R8 fate (Rh6 expression) in all R8s. The molecular nature of this signal is still unknown.[10,31]

In contrast to R7s that reuse Dve to differentiate subtypes, R8s utilize a very different mechanism relying on a bistable, double-negative feedback loop to interpret the signal from R7s and determine robust Rh expression. In R8s that lack signaling from R7s, the Hippo tumor-suppressor pathway induces the default **y**R8 fate including activation of Rh6 and repression of Rh5 (Fig. 4F). In **y**R8s, the FERM-domain protein Merlin (Mer) and the WW-domain protein Kibra (Kib) constitutively activate the core complex of the Ser/Thr Kinase Hippo (Hpo), the scaffolding protein Salvador (Sav), the cofactor Mob as tumor suppressor (Mats), and the Ser/Thr kinase Warts (Wts) to induce expression of Rh6 and repression of Rh5 (Fig. 4F). Control of the R8 subtype does not require the activity of the FERM-domain

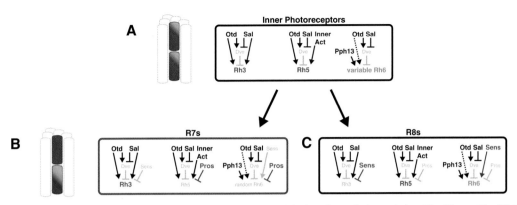

Figure 3. Binary switches limit Rh3 to R7s, and Rh5 and Rh6 to R8s. (A) Rh3, Rh5, and Rh6 are induced in all inner PRs. (B) Rh3 is expressed in R7s. Pros represses Rh5 and Rh6 in R7s, restricting their expression to R8s. (C) Rh5 and Rh6 are expressed in R8s. Sens induces robust expression of Rh6 in R8s. Sens represses Rh3 in R8s, restricting its expression to R7s.

protein Expanded (Ex) or the atypical cadherin fat that play critical roles in normal growth regulation, suggesting that the regulatory network has been altered for its new role in fate determination.[33,34]

pR7s signal to R8s to inhibit the Hippo tumor-suppressor pathway and induce **pR8** fate (Fig. 4E). Another growth regulator, the PH domain–containing protein Melted (Melt), acts in a bistable feedback loop with the Hippo pathway member, Wts, to control R8 subtype fate (Fig. 4E, F, and I). Upon signaling from **pR7s**, Melt expression is upregulated and Wts expression is downregulated. The loss of Wts causes expression of Rh5 and repression of Rh6. In the absence of the signal, Wts feeds back to maintain repression of Melt and ensure **y**R8 fate.[33,34]

The reimplementation of the Hippo growth pathway in R8 subtype specification is one of the most curious mechanisms utilized for fate determination in the fly eye.[33,34] In growth regulation, the Hippo pathway incorporates multiple negative feedback loops to ensure a balance between high and low proliferative states that are susceptible to regulation by multiple inputs. In contrast, in R8 subtype specification, the Hippo pathway is rewired and acts in a bistable, double-negative feedback loop with Melt to stably lock in R8 subtype fate. In this way, the Hippo pathway has been reimplemented, but altered, for its very different role in gene regulation in R8s. Though Melt is a key to this altered network topology, it is unclear how Melt functions to regulate the transcription of Wts and vice versa, since neither of these proteins are transcription factors.

Rhodopsin feeds back to maintain PR subtype fate

Many of the regulatory factors required for establishment of PR subtype fate are also needed to maintain proper Rh expression. However, an interesting feedback mechanism is dedicated specifically to maintaining Rh expression in exclusive subtypes in R8s: Rh6 feeds back to maintain repression of Rh5 in yR8s (Fig. 4G and H). This feedback is light-dependent yet is independent of the phototransduction signal transduction pathway.[35] It makes sense that this Rh feedback does not occur in R7s where coexpression occurs in specialized dorsal third ommatidia (see below).

Like in the fly color vision system, feedback from sensory receptors to maintain repression of other receptors is critical in the olfactory system of mice. In that case, each olfactory neuron selects expression of one olfactory receptor gene from a battery of approximately 1300 possibilities. The expressed receptor feeds back to prevent expression of the other receptor genes.[36–41] The fly color vision system incorporates a similar mechanism of receptor feedback; however it is specific to Rh6-mediated repression of Rh5 in yR8s. Since these feedback mechanisms do not rely on canonical phototransduction downstream of G protein–coupled receptor signaling, it will be interesting to uncover the novel regulatory pathways that maintain repression.

Regionally specified ommatidial subtypes

In addition to the randomly distributed **p** and **y** ommatidial subtypes, the retina contains two

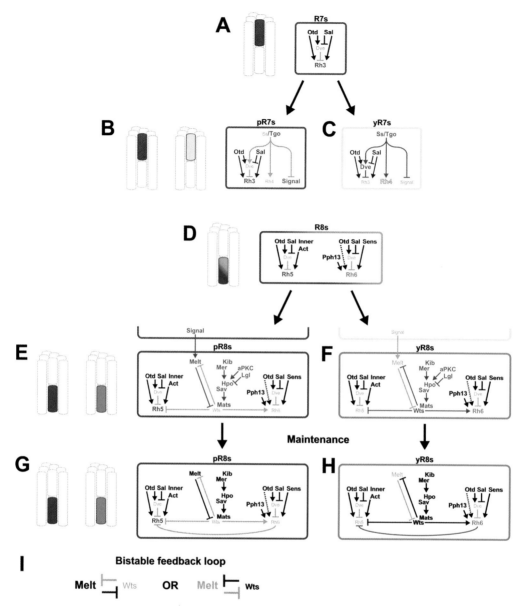

Figure 4. Control of Rhs in PR subtypes. (A) Rh3 is induced in all R7s. (B) In **pR7s** that lack Ss, default **pR7** fate is assumed including expression of Rh3 and the signal to R8. (C) In **yR7s**, Ss induces Dve to repress Rh3, activates Rh4, and represses the signal to R8. (D) Rh5 and Rh6 are induced in all R8s. (E) In **pR8s**, the signal from **pR7s** induces Melt expression that represses Wts expression, preventing Hippo pathway activity and allowing for Rh5 expression and loss of Rh6 expression. (F) In the absence of the signal from R7s, the Hippo pathway is active and induces default **yR8** fate including repression of Rh5 and expression of Rh6. (G) There is no Rh feedback maintenance in **pR8s**. (H) Rh6 acts to maintain repression of Rh5 in **yR8s**. (I) Melt and Wts interact in a double-negative bistable feedback loop to stabilize **pR8** and **yR8** fates.

specialized ommatidial subtypes that are determined by regional inputs. In the dorsal third of the retina, Rh3 is induced in Rh4-expressing **yR7s**, leading to coexpression of Rh3 and Rh4 in **yR7s** in this region (Fig. 1B and E). The transcription factors of the Iroquois complex (IroC) are specifically expressed in dorsal third R7s in adults and function to activate expression of Rh3 in **yR7s**. However, IroC is not adequate to induce Rh3 due to the presence of the Dve repressor in **yR7s**. To relieve this

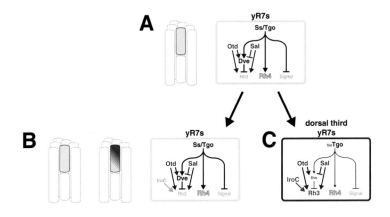

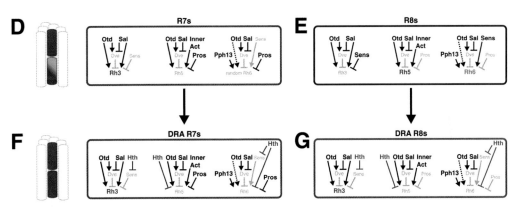

Figure 5. Regionally specified ommatidial subtypes. (A) In **y**R7s, Ss induces Rh4 and repression of Rh3 (via Dve) and the signal to R8. (B) In **y**R7s in the main part of the retina, Ss at high levels induces Rh4 and represses Rh3 (via Dve) and the signal to R8s. (C) In dorsal third **y**R7s, attenuation of Ss yields reduction in Dve levels allowing IroC-mediated activation of Rh3. (D) Rh3 is induced in all R7s. (E) Rh5 and Rh6 are induced in all R8s. (F) In DRA R7s, Rh3 is induced, and Rh5 and Rh6 are repressed (likely through the default R7 fate machinery). (G) In DRA R8s, Hth represses Sens to allow for Rh3 expression. Hth represses Rh5 through mechanisms independent of the known regulators. Hth represses Rh6 by repressing the Sens activator and through mechanisms independent of the known regulators.

repression, Ss levels are attenuated in the dorsal third, causing a decrease in Dve levels and allowing IroC-mediated activation of Rh3. Thus, Rh3 expression in Rh4-expressing dorsal third **y**R7s requires both activation by IroC and relief of repression by Dve (and indirectly Ss) (Fig. 5A–C).[15,18,32]

Specialized ommatidia are also found in the row at the dorsal rim of the retina (DRA). In this region, the homeodomain protein Homothorax (Hth) overrides the subtype specification program to repress Rh5 and Rh6 and induce Rh3 expression in both R7s and R8s. Hth likely allows Rh3 expression in R8s by repressing Sens, a repressor of Rh3. Hth appears to repress Rh5 and Rh6 by repressing Sens (an activator of Rh6) and by additional mechanisms that are independent of known regulators (Fig. 5D–G). Hth also induces changes in rhabdomere morphology, changing the size from the small inner PR size to the large outer PR size and the orientation from orthogonal to parallel for their specialized role in the detection of polarized light.[16,42,43]

The puzzle of randomness

The random mosaic of ommatidial subtypes is a beautiful feature of the fly color vision system, but why is it stochastic? Wouldn't the system be better served by an organized pattern? Lineage- and signaling-based deterministic developmental programs are the best-understood mechanisms to specify cell fates in a reproducible fashion. It appears that stochastic cell fate specification is an important yet underappreciated and poorly understood mechanism to diversify cell fates during development. In bacteria, stochastic gene expression is utilized to generate random, transient subpopulations of cells as a means of bet-hedging against drastic changes to harsh environments (e.g., persister cells and competence for DNA uptake).[44–51] In vertebrates, stochastic mechanisms are utilized for a variety of stable cell fate phenomena including olfactory receptor and color opsin choice,[36–41,52–80] motor neuron subtype specification,[81–89] protocadherin expression,[90–96] B cell development,[97,98] and stem cell differentiation.[6,99–101]

Though stochastic gene expression processes appear to be critical for diversifying cell fates across species, the reasons for using such mechanisms remain mysterious. One utilitarian hypothesis is that these development phenomena do not require precise spatial patterning but simply need specific proportions of cell fates that are roughly evenly distributed across a tissue. In this way, stochastic cell fate specification mechanisms can be random yet robust in that they generate highly reproducible frequencies of cell fates. The fly eye is a perfect example of this robustness of cell fate proportions where the ratio of **p** to **y** fates is reproducibly 1:2 with variance across retinas similar to what is expected for a process controlled by a single stochastic event (i.e., the stochastic on/off regulation of Ss). Though the proportions of cell fates generated by stochastic cell fate specification have not been characterized for many cases, it is obvious that significant deviations from a norm would be highly deleterious for some phenomena. For example, the generation of motor neuron subtypes requires stochastic specification coupled with cadherin-mediated migration to create reproducible motor neuron pools. Dramatic changes in the frequency of cell fates generate underrepresented motor pools presumably causing defects in movement.[81–83] In contrast, some stochastic systems appear to tolerate deviations in cell fate proportions across individuals of a population. One case is the human color vision system in which the ratio of color opsins in cone photoreceptors (red/green/blue) can vary greatly among individuals.[61] Thus, key questions concern not only the actual molecular mechanisms controlling these stochastic processes but also the ways that these systems generate reproducible proportions of cell fates or allow for high variation.

Though the 1:2 proportion of PR fates in the fly eye appears to be conserved across higher dipterans, the importance of this ratio for the color vision system is unclear. Genetic ablation of specific rhodopsins dramatically alters the fly's innate preference for specific colors.[102,103] However, whether the proportion of **p** versus **y** cell fates dramatically affects the fly's capacity to respond to color is not clear. Additionally, though innate spectral preference is easily observable, true color vision in flies (i.e., response to distinct wavelengths of light independent of intensity) has not yet been clearly demonstrated. Comparative studies with other insect species that have reproducible patterns of ommatidial subtypes may provide insight. The ommatidial subtypes of mosquito species like *Aedes aegypti* and *Anopheles gambiae* occur in dorsal/ventral patterns[104] whereas the subtypes of some fly species like *Dolichopodidae* appear to be generated in dorsal to ventral stripes.[105–108] Study of these species may reveal the impact of stochastic versus deterministic ommatidial subtype patterning on behavior. It will be important to determine whether species that display deterministic subtype patterning utilize the same molecular machinery used for stochastic specification of fates in *Drosophila*, or have evolved unique mechanisms. Dve is expressed in outer PRs of mosquitoes similar to *Drosophila*, suggesting that divergent species displaying vastly different Rh expression patterns share some of the same regulatory molecular machinery.[18] If the molecular players are shared among these species, it will be exciting to discern how the regulation of these genes has been altered for their new roles. For example, *ss* may have been converted from a randomly expressed gene to a reproducibly expressed gene.

Conclusion

Though studied for over 100 years, the fly eye paradigm continues to offer insights into the molecular mechanisms controlling cell fate determination. Examination of the random mosaic of ommatidial subtypes determined by expression of color-detecting Rhs in the inner PRs has inspired new ideas about how gene regulatory mechanisms are layered to dictate robust cell fate specification. Despite a great deal of progress, there are still numerous questions left to be answered. The most pressing questions concern the control of the stochastic expression of *ss*. What mechanisms determine whether *ss* will be expressed or not? What mechanistic properties will be shared among stochastic cell fate specification phenomena? How is the 1:2 ratio established and why is it important? Answers to these questions will provide fundamental understanding into the mysterious process of stochastic fate determination, which will have a broad impact on biological phenomena from the development of sensory and motor systems to the differentiation of stem cells.

Acknowledgments

Thank you to Lionel Christiaen, Claude Desplan, David Jukam, and Daniel Vasiliauskas for helpful comments. R.J.J. was supported by a Jane Coffin Childs Memorial Fund for Medical Research postdoctoral fellowship.

Conflicts of interest

The author declares no conflicts of interest.

References

1. Inwood, S. 2002. *The Man Who Knew Too Much: The Strange and Inventive Life of Robert Hooke, 1635–1703.* London: Pan Macmillan.
2. Wolff, T. & D.F. Ready. 1991. The beginning of pattern formation in the Drosophila compound eye: the morphogenetic furrow and the second mitotic wave. *Development* **113:** 841–850.
3. Hardie, R.C. 1985. "Functional organization of the fly retina." In *Progress in Sensory Physiology.* H. Autrum, D. Ottoson, E.R. Perl, R.F. Schmidt, H. Shimazu & W.D. Willis, Eds.: 1–79. Berlin: Springer.
4. O'Tousa, J.E. *et al.* 1985. The Drosophila ninaE gene encodes an opsin. *Cell* **40:** 839–850.
5. Zuker, C.S., A.F. Cowman & G.M. Rubin. 1985. Isolation and structure of a rhodopsin gene from D. melanogaster. *Cell* **40:** 851–858.
6. Johnston, R.J., Jr. & C. Desplan. 2008. Stochastic neuronal cell fate choices. *Curr. Opin. Neurobiol.* **18:** 20–27.
7. Johnston, R.J., Jr. & C. Desplan. 2010. Preview: a penetrating look at stochasticity in development. *Cell* **140:** 610–612.
8. Bell, M.L., J.B. Earl & S.G. Britt. 2007. Two types of Drosophila R7 photoreceptor cells are arranged randomly: a model for stochastic cell-fate determination. *J. Comp. Neurol.* **502:** 75–85.
9. Chou, W.H. *et al.* 1996. Identification of a novel Drosophila opsin reveals specific patterning of the R7 and R8 photoreceptor cells. *Neuron* **17:** 1101–1115.
10. Chou, W.H. *et al.* 1999. Patterning of the R7 and R8 photoreceptor cells of Drosophila: evidence for induced and default cell-fate specification. *Development* **126:** 607–616.
11. Montell, C., K. Jones, C. Zuker & G. Rubin. 1987. A second opsin gene expressed in the ultraviolet-sensitive R7 photoreceptor cells of Drosophila melanogaster. *J. Neurosci.* **7:** 1558–1566.
12. Papatsenko, D., G. Sheng & C. Desplan. 1997. A new rhodopsin in R8 photoreceptors of Drosophila: evidence for coordinate expression with Rh3 in R7 cells. *Development* **124:** 1665–1673.
13. Zuker, C.S., C. Montell, K. Jones, *et al.* 1987. A rhodopsin gene expressed in photoreceptor cell R7 of the Drosophila eye: homologies with other signal-transducing molecules. *J. Neurosci.* **7:** 1550–1557.
14. Fortini, M.E. & G.M. Rubin. 1990. Analysis of cis-acting requirements of the Rh3 and Rh4 genes reveals a bipartite organization to rhodopsin promoters in Drosophila melanogaster. *Genes Dev.* **4:** 444–463.
15. Mazzoni, E.O. *et al.* 2008. Iroquois complex genes induce co-expression of rhodopsins in Drosophila. *PLoS Biol.* **6:** e97.
16. Wernet, M.F. *et al.* 2003. Homothorax switches function of Drosophila photoreceptors from color to polarized light sensors. *Cell* **115:** 267–279.
17. Tomlinson, A. 2003. Patterning the peripheral retina of the fly: decoding a gradient. *Dev. Cell* **5:** 799–809.
18. Johnston, R.J., Jr. *et al.* 2011. Interlocked feedforward loops control cell-type-specific rhodopsin expression in the Drosophila eye. *Cell* **145:** 956–968.
19. Mishra, M. *et al.* 2010. Pph13 and orthodenticle define a dual regulatory pathway for photoreceptor cell morphogenesis and function. *Development* **137:** 2895–2904.
20. Tahayato, A. *et al.* 2003. Otd/Crx, a dual regulator for the specification of ommatidia subtypes in the Drosophila retina. *Dev. Cell* **5:** 391–402.
21. Sood, P., R.J. Johnston, Jr. & E. Kussell. 2012. Stochastic de-repression of Rhodopsins in single photoreceptors of the fly retina. *PLoS Comput. Biol.* **8:** e1002357.
22. Alon, U. 2007. Network motifs: theory and experimental approaches. *Nat. Rev. Genet.* **8:** 450–461.
23. Goentoro, L., O. Shoval, M.W. Kirschner & U. Alon. 2009. The incoherent feedforward loop can provide fold-change detection in gene regulation. *Mol. Cell* **36:** 894–899.
24. Kaplan, S., A. Bren, E. Dekel & U. Alon. 2008. The incoherent feed-forward loop can generate non-monotonic input functions for genes. *Mol. Syst. Biol.* **4:** 203.
25. Kim, D., Y.K. Kwon & K.H. Cho. 2008. The biphasic behavior of incoherent feed-forward loops in biomolecular regulatory networks. *Bioessays* **30:** 1204–1211.
26. Mangan, S. & U. Alon. 2003. Structure and function of the feed-forward loop network motif. *Proc. Natl. Acad. Sci. USA* **100:** 11980–11985.
27. Mollereau, B. *et al.* 2001. Two-step process for photoreceptor formation in Drosophila. *Nature* **412:** 911–913.
28. Xie, B., M. Charlton-Perkins, E. McDonald, *et al.* 2007. Senseless functions as a molecular switch for color photoreceptor differentiation in Drosophila. *Development* **134:** 4243–4253.

29. Kalir, S., S. Mangan & U. Alon. 2005. A coherent feed-forward loop with a SUM input function prolongs flagella expression in *Escherichia coli*. *Mol. Syst. Biol.* **1:** 0006.

30. Cook, T., F. Pichaud, R. Sonneville, *et al.* 2003. Distinction between color photoreceptor cell fates is controlled by Prospero in Drosophila. *Dev. Cell* **4:** 853–864.

31. Wernet, M.F. *et al.* 2006. Stochastic spineless expression creates the retinal mosaic for colour vision. *Nature* **440:** 174–180.

32. Thanawala, S.U., J. Rister, G.W. Goldberg, *et al.* 2013. Regional modulation of a stochastically expressed factor determines photoreceptor subtypes in the Drosophila retina. *Dev. Cell* **25:** 93–105.

33. Jukam, D. & C. Desplan. 2011. Binary regulation of Hippo pathway by Merlin/NF2, Kibra, Lgl, and Melted specifies and maintains postmitotic neuronal fate. *Dev. Cell* **21:** 874–887.

34. Mikeladze-Dvali, T. *et al.* 2005. The growth regulators warts/lats and melted interact in a bistable loop to specify opposite fates in Drosophila R8 photoreceptors. *Cell* **122:** 775–787.

35. Vasiliauskas, D. *et al.* 2011. Feedback from rhodopsin controls rhodopsin exclusion in Drosophila photoreceptors. *Nature* **479:** 108–112.

36. Buck, L. & R. Axel. 1991. A novel multigene family may encode odorant receptors: a molecular basis for odor recognition. *Cell* **65:** 175–187.

37. Godfrey, P.A., B. Malnic & L.B. Buck. 2004. The mouse olfactory receptor gene family. *Proc. Natl. Acad. Sci. USA* **101:** 2156–2161.

38. Lewcock, J.W. & R.R. Reed. 2004. A feedback mechanism regulates monoallelic odorant receptor expression. *Proc. Natl. Acad. Sci. USA* **101:** 1069–1074.

39. Serizawa, S. *et al.* 2003. Negative feedback regulation ensures the one receptor-one olfactory neuron rule in mouse. *Science* **302:** 2088–2094.

40. Shykind, B.M. *et al.* 2004. Gene switching and the stability of odorant receptor gene choice. *Cell* **117:** 801–815.

41. Zhang, X. & S. Firestein. 2002. The olfactory receptor gene superfamily of the mouse. *Nat. Neurosci.* **5:** 124–133.

42. Weir, P.T. & M.H. Dickinson. 2012. Flying Drosophila orient to sky polarization. *Curr. Biol.* **22:** 21–27.

43. Wernet, M.F. *et al.* 2012. Genetic dissection reveals two separate retinal substrates for polarization vision in Drosophila. *Curr. Biol.* **22:** 12–20.

44. Balaban, N.Q., J. Merrin, R. Chait, *et al.* 2004. Bacterial persistence as a phenotypic switch. *Science* **305:** 1622–1625.

45. Kussell, E. & S. Leibler. 2005. Phenotypic diversity, population growth, and information in fluctuating environments. *Science* **309:** 2075–2078.

46. Cagatay, T., M. Turcotte, M.B. Elowitz, *et al.* 2009. Architecture-dependent noise discriminates functionally analogous differentiation circuits. *Cell* **139:** 512–522.

47. Dubnau, D. 1999. DNA uptake in bacteria. *Annu. Rev. Microbiol.* **53:** 217–244.

48. Maamar, H. & D. Dubnau. 2005. Bistability in the Bacillus subtilis K-state (competence) system requires a positive feedback loop. *Mol. Microbiol.* **56:** 615–624.

49. Maamar, H., A. Raj & D. Dubnau. 2007. Noise in gene expression determines cell fate in Bacillus subtilis. *Science* **317:** 526–529.

50. Suel, G.M., J. Garcia-Ojalvo, L.M. Liberman & M.B. Elowitz. 2006. An excitable gene regulatory circuit induces transient cellular differentiation. *Nature* **440:** 545–550.

51. Suel, G.M., R.P. Kulkarni, J. Dworkin, *et al.* 2007. Tunability and noise dependence in differentiation dynamics. *Science* **315:** 1716–1719.

52. Hunt, D.M. *et al.* 1998. Molecular evolution of trichromacy in primates. *Vision Res.* **38:** 3299–3306.

53. Jacobs, G.H. & J. Nathans. 2009. The evolution of primate color vision. *Sci. Am.* **300:** 56–63.

54. Jacobs, G.H., M. Neitz, J.F. Deegan & J. Neitz. 1996. Trichromatic colour vision in New World monkeys. *Nature* **382:** 156–158.

55. Jacobs, G.H., G.A. Williams, H. Cahill & J. Nathans. 2007. Emergence of novel color vision in mice engineered to express a human cone photopigment. *Science* **315:** 1723–1725.

56. Kainz, P.M., J. Neitz & M. Neitz. 1998. Recent evolution of uniform trichromacy in a New World monkey. *Vision Res.* **38:** 3315–3320.

57. Nathans, J. 1999. The evolution and physiology of human color vision: insights from molecular genetic studies of visual pigments. *Neuron* **24:** 299–312.

58. Nathans, J. *et al.* 1989. Molecular genetics of human blue cone monochromacy. *Science* **245:** 831–838.

59. Nathans, J., T.P. Piantanida, R.L. Eddy, *et al.* 1986. Molecular genetics of inherited variation in human color vision. *Science* **232:** 203–210.

60. Nathans, J., D. Thomas & D.S. Hogness. 1986. Molecular genetics of human color vision: the genes encoding blue, green, and red pigments. *Science* **232:** 193–202.

61. Roorda, A. & D.R. Williams. 1999. The arrangement of the three cone classes in the living human eye. *Nature* **397:** 520–522.

62. Smallwood, P.M., Y. Wang & J. Nathans. 2002. Role of a locus control region in the mutually exclusive expression of human red and green cone pigment genes. *Proc. Natl. Acad. Sci. USA* **99:** 1008–1011.

63. Wang, Y. *et al.* 1992. A locus control region adjacent to the human red and green visual pigment genes. *Neuron* **9:** 429–440.

64. Wang, Y. *et al.* 1999. Mutually exclusive expression of human red and green visual pigment-reporter transgenes occurs at high frequency in murine cone photoreceptors. *Proc. Natl. Acad. Sci. USA* **96:** 5251–5256.

65. Chess, A., I. Simon, H. Cedar & R. Axel. 1994. Allelic inactivation regulates olfactory receptor gene expression. *Cell* **78:** 823–834.

66. Eggan, K. *et al.* 2004. Mice cloned from olfactory sensory neurons. *Nature* **428:** 44–49.

67. Fuss, S.H., M. Omura & P. Mombaerts. 2007. Local and cis effects of the H element on expression of odorant receptor genes in mouse. *Cell* **130:** 373–384.

68. Khan, M., E. Vaes & P. Mombaerts. 2011. Regulation of the probability of mouse odorant receptor gene choice. *Cell* **147:** 907–921.

69. Li, J., T. Ishii, P. Feinstein & P. Mombaerts. 2004. Odorant receptor gene choice is reset by nuclear transfer from mouse olfactory sensory neurons. *Nature* **428:** 393–399.

70. Lomvardas, S. *et al.* 2006. Interchromosomal interactions and olfactory receptor choice. *Cell* **126:** 403–413.

71. Magklara, A. *et al.* 2011. An epigenetic signature for monoallelic olfactory receptor expression. *Cell* **145:** 555–570.

72. Miyamichi, K., S. Serizawa, H.M. Kimura & H. Sakano. 2005. Continuous and overlapping expression domains of odorant receptor genes in the olfactory epithelium determine the dorsal/ventral positioning of glomeruli in the olfactory bulb. *J. Neurosci.* **25:** 3586–3592.

73. Nguyen, M.Q., Z. Zhou, C.A. Marks, *et al.* 2007. Prominent roles for odorant receptor coding sequences in allelic exclusion. *Cell* **131:** 1009–1017.

74. Nishizumi, H., K. Kumasaka, N. Inoue, *et al.* 2007. Deletion of the core-H region in mice abolishes the expression of three proximal odorant receptor genes in cis. *Proc. Natl. Acad. Sci. USA* **104:** 20067–20072.

75. Ressler, K.J., S.L. Sullivan & L.B. Buck. 1993. A zonal organization of odorant receptor gene expression in the olfactory epithelium. *Cell* **73:** 597–609.

76. Rothman, A., P. Feinstein, J. Hirota & P. Mombaerts. 2005. The promoter of the mouse odorant receptor gene M71. *Mol. Cell Neurosci.* **28:** 535–546.

77. Tian, H. & M. Ma. 2008. Activity plays a role in eliminating olfactory sensory neurons expressing multiple odorant receptors in the mouse septal organ. *Mol. Cell Neurosci.* **38:** 484–488.

78. Vassalli, A., A. Rothman, P. Feinstein, *et al.* 2002. Minigenes impart odorant receptor-specific axon guidance in the olfactory bulb. *Neuron* **35:** 681–696.

79. Vassar, R., J. Ngai & R. Axel. 1993. Spatial segregation of odorant receptor expression in the mammalian olfactory epithelium. *Cell* **74:** 309–318.

80. Clowney, E. J. *et al.* 2012. Nuclear aggregation of olfactory receptor genes governs their monogenic expression. *Cell* **151:** 724–737.

81. Dasen, J.S. & T.M. Jessell. 2009. Hox networks and the origins of motor neuron diversity. *Curr. Top. Dev. Biol.* **88:** 169–200.

82. Dasen, J.S., J.P. Liu & T.M. Jessell. 2003. Motor neuron columnar fate imposed by sequential phases of Hox-c activity. *Nature* **425:** 926–933.

83. Dasen, J.S., B.C. Tice, S. Brenner-Morton & T.M. Jessell. 2005. A Hox regulatory network establishes motor neuron pool identity and target-muscle connectivity. *Cell* **123:** 477–491.

84. Helmbacher, F. *et al.* 2003. Met signaling is required for recruitment of motor neurons to PEA3-positive motor pools. *Neuron* **39:** 767–777.

85. Jessell, T.M. 2000. Neuronal specification in the spinal cord: inductive signals and transcriptional codes. *Nat. Rev. Genet.* **1:** 20–29.

86. Lin, J.H. *et al.* 1998. Functionally related motor neuron pool and muscle sensory afferent subtypes defined by coordinate ETS gene expression. *Cell* **95:** 393–407.

87. Livet, J. *et al.* 2002. ETS gene Pea3 controls the central position and terminal arborization of specific motor neuron pools. *Neuron* **35:** 877–892.

88. Price, S.R., N.V. De Marco Garcia, B. Ranscht & T.M. Jessell. 2002. Regulation of motor neuron pool sorting by differential expression of type II cadherins. *Cell* **109:** 205–216.

89. Theriault, F.M., P. Roy & S. Stifani. 2004. AML1/Runx1 is important for the development of hindbrain cholinergic branchiovisceral motor neurons and selected cranial sensory neurons. *Proc. Natl. Acad. Sci. USA* **101:** 10343–10348.

90. Guo, Y. *et al.* 2012. CTCF/cohesin-mediated DNA looping is required for protocadherin alpha promoter choice. *Proc. Natl. Acad. Sci. USA* **109:** 21081–21086.

91. Li, Y. *et al.* 2012. Molecular and functional interaction between protocadherin-gamma C5 and GABAA receptors. *J. Neurosci.* **32:** 11780–11797.

92. Chen, W.V. *et al.* 2012. Functional significance of isoform diversification in the protocadherin gamma gene cluster. *Neuron* **75:** 402–409.

93. Monahan, K. *et al.* 2012. Role of CCCTC binding factor (CTCF) and cohesin in the generation of single-cell diversity of protocadherin-alpha gene expression. *Proc. Natl. Acad. Sci. USA* **109:** 9125–9130.

94. Kehayova, P., K. Monahan, W. Chen & T. Maniatis. 2011. Regulatory elements required for the activation and repression of the protocadherin-alpha gene cluster. *Proc. Natl. Acad. Sci. USA* **108:** 17195–17200.

95. Ribich, S., B. Tasic & T. Maniatis. 2006. Identification of long-range regulatory elements in the protocadherin-alpha gene cluster. *Proc. Natl. Acad. Sci. USA* **103:** 19719–19724.

96. Tasic, B. *et al.* 2002. Promoter choice determines splice site selection in protocadherin alpha and gamma pre-mRNA splicing. *Mol. Cell* **10:** 21–33.

97. Tarlinton, D. 2012. B-cell differentiation: instructive one day, stochastic the next. *Curr. Biol.* **22:** R235–R237.

98. Duffy, K.R. *et al.* 2012. Activation-induced B cell fates are selected by intracellular stochastic competition. *Science* **335:** 338–341.

99. He, J. *et al.* 2012. How variable clones build an invariant retina. *Neuron* **75:** 786–798.

100. Gomes, F.L. *et al.* 2011. Reconstruction of rat retinal progenitor cell lineages in vitro reveals a surprising degree of stochasticity in cell fate decisions. *Development* **138:** 227–235.

101. Johnston, R.J., Jr. & C. Desplan. 2010. Stochastic mechanisms of cell fate specification that yield random or robust outcomes. *Annu. Rev. Cell. Dev. Biol.* **26:** 689–719.

102. Yamaguchi, S., C. Desplan & M. Heisenberg. 2010. Contribution of photoreceptor subtypes to spectral wavelength preference in Drosophila. *Proc. Natl. Acad. Sci. USA* **107:** 5634–5639.

103. Yamaguchi, S., R. Wolf, C. Desplan & M. Heisenberg. 2008. Motion vision is independent of color in Drosophila. *Proc. Natl. Acad. Sci. USA* **105:** 4910–4915.

104. Hu, X. *et al.* 2009. Patterned rhodopsin expression in R7 photoreceptors of mosquito retina: implications for species-specific behavior. *J. Comp. Neurol.* **516:** 334–342.

105. Trujillo-Cenoz, O. & G.D. Bernard. 1972. Some aspects of the retinal organization of Sympycnus linetaus

Loew (Diptera, Dolichopodidae). *J. Ultrastruc. Res.* **38:** 149–160.

106. Tanaka, G., A.R. Parker, D.J. Siveter, *et al.* 2009. An exceptionally well-preserved Eocene dolichopodid fly eye: function and evolutionary significance. *Proc. Biol. Sci.* **276:** 1015–1019.

107. Lunau, K. & H. Knuettel. 1995. Vision through colored eyes. *Naturwissenschaften* **82:** 432–434.

108. Knuettel, H. & K. Lanau. 1997. Farbige Augen bei Insekten—Mitteilungen der Deutschen Gesellschaft fuer allgemeine und angewandte. *Entomologie* **11:** 587–590.

Ann. N.Y. Acad. Sci. ISSN 0077-8923

ANNALS OF THE NEW YORK ACADEMY OF SCIENCES
Issue: *Blavatnik Awards for Young Scientists 2012*

Evidence for wave heating in the solar corona

Michael Hahn

Columbia Astrophysics Laboratory, Columbia University, New York, New York

Address for correspondence: Michael Hahn, Columbia Astrophysics Laboratory, Columbia University, MC 5247, 550 West 120th Street, New York, NY 10027. mhahn@astro.columbia.edu

The temperature of the Sun increases over a short distance from a few thousand degrees in the photosphere to over a million degrees in the corona. To understand coronal heating is one of the major problems in astrophysics. There is general agreement that the energy source is convective motion in and below the photosphere. It remains to determine how this mechanical energy is transported outward into the corona and then deposited as heat. Two classes of models have been proposed, namely those that rely on magnetic reconnection and those that rely on waves, particularly Alfvén waves. There is increasing evidence that waves are ubiquitous in the corona. However, a difficulty for wave-driven models has been that most theories predict Alfvén waves to be undamped in the corona, and therefore they cannot dissipate their energy into heat. Our research has shown unambiguous observational evidence that the waves do damp at sufficiently low heights in the corona to be important for coronal heating.

Keywords: solar physics; coronal heating; Sun; plasma waves

Introduction

The corona is the hot diffuse outer atmosphere of the Sun that is visible during a solar eclipse. It extends from slightly above the photosphere, the visible surface of the Sun, far into space. The corona is the subject of one of the major problems in astrophysics, known as the coronal heating problem. In the core of the Sun, where nuclear fusion takes place, the temperature is about 15×10^6 K. Moving outward from the core, the temperature drops to a minimum of about 4400 K at the photosphere.[1] About 70 years ago emission lines from highly ionized iron were identified in the coronal spectrum, which proved that the temperature of the corona is over 10^6 K, more than 100 times hotter than the photosphere.[2,3]

The thermal pressure of the hot corona causes it to expand into space, forming the continuous outflow of charged particles known as the solar wind.[4,5] The interaction between the solar wind and the Earth's magnetosphere produces the aurorae. During periods of high activity, solar storms, such as coronal mass ejections, are carried within the solar wind to the Earth where they can damage satel-

lites, and disrupt communications and electrical grids.[6,7]

So far there has been no definitive explanation for the high temperature of the corona and the consequent acceleration of the solar wind. Two main models have been proposed—one in which heating occurs by reconnection of magnetic field lines, and another in which the heating is accomplished through dissipation of plasma waves. In both cases the energy source is the mechanical energy seen as turbulent convective motion in the photosphere. This energy is transmitted to the corona by the magnetic field, which permeates the solar atmosphere from the photosphere to the corona. Because of the solar plasma's high conductivity, the magnetic field lines move with the fluid. As a result, the churning of the plasma in the photosphere tangles the magnetic field lines.

Magnetic reconnection occurs when there is a large gradient in the magnetic field. Such gradients are built up by the photospheric motions, for example, by bringing close together two magnetic field lines of opposite polarity. When the gradient is very large the magnetic field lines are no longer constrained to move with the fluid and instead

doi: 10.1111/nyas.12066

the field lines diffuse through the region of strong gradients.[8] As the magnetic field diffuses the energy stored in the stressed fields is dissipated. Much of this energy goes into electric currents, which heat the plasma through the plasma resistivity, known as Ohmic heating, and some energy may also be carried away by plasma waves.[9,10] Magnetic reconnection is known to be the cause of solar flares and coronal mass ejections. The difficulty for reconnection-driven coronal heating models is that the corona appears to be in a steady state, whereas reconnection heats the plasma in bursts. In order for reconnection to appear as steady heating there must be a large number of small reconnection events happening. Because small reconnection events are hard to detect it is not known whether this scenario actually occurs on the Sun.[11–13]

The jostling of the magnetic field also drives waves that can propagate into the corona. The type of waves most likely to be responsible for coronal heating are known as Alfvén waves. These waves travel along the magnetic field line in a manner that is analogous to waves travelling along a taught string. Such waves have been observed in the Sun in the chromosphere,[14] the low corona,[15,16] and the solar wind.[17] For waves to be responsible for coronal heating the energy they carry must be dissipated into heat below about 2 $R_\odot$ (1 $R_\odot = 6.96 \times 10^8$ m). This is because at heights above two solar radii the plasma density is too low to conduct the heat down-

ward and bring the corona to the relatively uniform temperature that is observed. A problem for wave-driven models has been that Alfvén wave damping is predicted to require length scales of more than 3 $R_\odot$.[18–20]

Analysis and results

We used data from the Extreme Ultraviolet Imaging Spectrometer (EIS)[21] on the Hinode satellite[22] to find evidence of wave damping in the corona.[23] The EIS instrument observes spectral lines in the wavelength range of 171–211 Å and 245–291 Å with a spectral resolution of 0.022 Å/pixel. A portion of the EIS spectrum is shown in Figure 1. In the EIS wavelength range, the coronal spectrum consists mainly of lines from highly ionized iron and silicon, with some lines from other elements.[24,25]

We studied an observation of a polar coronal hole. Coronal holes are regions where the magnetic field lines are open, that is, they extend far out into space.[19,26] During solar minimum, coronal holes are found at the Sun's polar regions. Coronal holes appear dark at ultraviolet wavelengths because they are less dense and have slightly cooler temperatures than other regions of the corona. They are interesting because plasma can stream out along their open field lines forming the fast solar wind.[27,28] Figure 2 indicates the pointings of the EIS spectrometer slit for our data, which covered the height range from below the solar limb up to 1.4 $R_\odot$. The background image

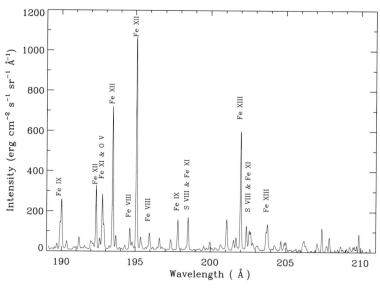

Figure 1. A portion of the EIS spectrum with labels on selected lines.

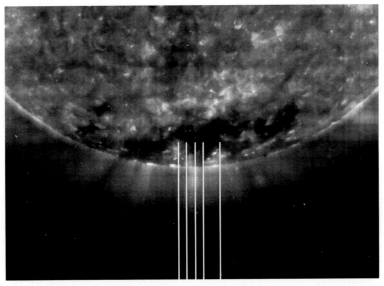

Figure 2. Image of the Sun taken with the Extreme Ultraviolet Imaging Telescope on the *SOHO* spacecraft. The telescope filter passes light near 171 Å and is sensitive to emission from material having electron temperatures of about 8 × 10⁵ K. The white lines indicate the pointing of the EIS spectrometer for our data.

in Figure 2 was taken with the Extreme Ultraviolet Imaging Telescope[29] on the *SOHO* spacecraft[30] and shows material at an electron temperature (T_e) of about 8×10^5 K.[31]

The sloshing of the plasma caused by the waves as they pass through it causes a nonthermal Doppler broadening of the emission lines.[32–35] Thus, waves can be detected by measuring emission line widths. The width of a spectral line $\Delta\lambda$ is given by

$$\Delta\lambda = \sqrt{\frac{\lambda}{c}\left(\frac{2kT_i}{M} + v_{nt}^2\right)}, \quad (1)$$

where λ is the rest wavelength of the line, c is the speed of light, M is the ion mass, k is the Boltzmann constant, T_i is the ion temperature, and v_{nt} is the nonthermal velocity. The wave amplitude is proportional to v_{nt}.

If there is no damping, then the variation with height of the wave amplitude, and hence v_{nt}, can be predicted from energy conservation. The energy flux carried by harmonic plane Alfvén waves is $F = 2\rho v_{nt}^2 V_A$. Here, ρ is the mass density and the speed of the waves is $V_A = B/\sqrt{4\pi\rho}$, where B is the magnetic field strength. If wave energy is conserved, then the flux $F A$ crossing area A is constant. The waves travel along magnetic field lines, which separate from one another as the magnetic field

strength drops. Consequently, $B A$ is also constant. Therefore, one predicts $v_{nt} \propto \rho^{-1/4}$ for undamped waves.[36–38] We expect that since density decreases with height, the wave amplitude will increase with height if the waves are undamped.

We have measured the line width as a function of height to look for damping. Here, we will express the line width in terms of an effective velocity,

$$v_{eff} = \sqrt{\frac{2kT_i}{M} + v_{nt}^2}. \quad (2)$$

Since v_{eff} also includes the ion temperature, there is ambiguity about whether a decrease in v_{eff} represents a decrease in T_i or v_{nt}. However, other observations have shown that T_i increases with height, and so any decrease in v_{eff} can be attributed to v_{nt}.[39]

A limitation in previous attempts to measure wave damping using this method has been instrumental scattered light. This is an effect in which light from the bright solar disk is reflected off the surface roughness of the optics within the spectrometer and becomes superimposed on the off-disk data. Line widths in disk spectra are narrower than in the off-disk data. This is because the light comes from lower layers of the atmosphere where the plasma is cooler and denser. If scattered light contamination is large, then the line widths will appear to be narrower.

Previous observations of line narrowing have been ascribed to this instrumental effect.[40,41]

To correct for this effect we estimated the intensity of scattered light and then subtracted it from our data. We used the intensity of a He II (i.e., He$^+$) line as a proxy for the scattered light. He II is abundant when T_e is below about 10^5 K, but because the corona is much hotter than this the He II should be mostly ionized into He III. Thus, we expect that the observed He II emission at large heights is not real, but is instead due to scattered light. This allowed us to infer the scattered light intensity. The scattered light fraction of the total intensity increased with height as the intensity of the real emission decreases. At the largest heights we studied, scattered light contributed between 10% and 45% of the total intensity. Using the He II emission as a proxy, we were able to subtract the scattered light from our spectra and determine the line widths for the spectrum of real emission from the corona.[23]

Figure 3 shows measured line widths as a function of height for several spectral lines from Si x, Fe IX, and Fe x (i.e., Si^{9+}, Fe^{8+}, and Fe^{9+}, respectively). The line widths initially follow the $\rho^{-1/4}$ dependence expected for undamped waves, but are damped starting at about 1.2 $R_\odot$. These results have been confirmed by another group studying a different observation.[42]

Implications for coronal heating

From our measurements we can estimate the amount of energy deposited in the corona by the waves. As stated earlier, the energy flux for harmonic plane Alfvén waves is $F = 2\rho v_{\rm nt}^2 V_A$. From these measurements, we determined the energy flux at two heights in the corona, one below where we observe the damping to start, 1.1 $R_\odot$, and one above the start of the damping, 1.3 $R_\odot$. To do so we found the density ρ, the magnetic field strength B, and estimated $v_{\rm nt}$ from $v_{\rm eff}$.

We determined the electron density (n_e) using the intensity ratio of two lines, Fe IX 188.50 Å and 189.94 Å, which has a known dependence on n_e. The mass density is then $\rho \approx m_p n_e$, where m_p is the proton mass. The magnetic field strength in polar coronal holes has been measured to be about 7 G at 1 $R_\odot$.[43] For larger heights we can estimate B if we know how the area $A(R)$ expands with radius. We took these area expansion factors $A(R)/A(R_\odot)$ from an empirical model.[44] The nonthermal velocity cannot be directly determined since we do not have an independent measurement of T_i. However, other measurements have shown that $T_i > T_e \approx 1 \times 10^6$ K.[45] Therefore, we found an upper bound for $v_{\rm nt}$ by assuming $T_i = T_e$ in Eq. (2). This leads to

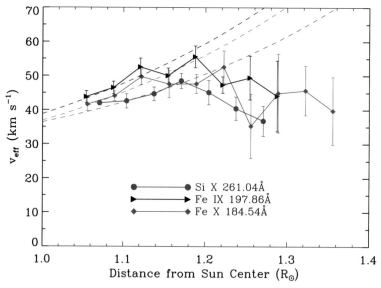

Figure 3. Symbols show line widths, $v_{\rm eff}$, as a function of height, in units of solar radii, for lines from three ions, Si x, Fe IX, and Fe x. The dashed lines show the $\rho^{-1/4}$ trend expected for the line widths if there were no damping. The deviation of the measurements from these trends indicates wave damping starting at about 1.2 $R_\odot$.

$v_{nt} = 48.6$ km/s at $1.1\ R_\odot$ and 46.1 km/s at $1.3\ R_\odot$. Putting all these values together, we found that the energy fluxes are $F \leq 7.0 \times 10^5$ erg/cm^2/s at $1.1\ R_\odot$ and $F \leq 1.1 \times 10^5$ erg/cm^2/s at $1.3\ R_\odot$. These values are upper bounds because the thermal broadening is not independently determined. In addition, we have estimated F using the the expression for plane waves, which is only approximate for the real structured solar corona.

These energy fluxes indicate that up to 70% of the energy initially in the waves is damped away between $1.1\ R_\odot$ and $1.3\ R_\odot$. The separation of the magnetic field lines implies that the area expands by a factor of two between $1.1\ R_\odot$ and $1.3\ R_\odot$. Since FA is constant and we observe an upper limit of $F(1.1 R_\odot) = 7 \times 10^5$ erg/cm^2/s, we expect $F(1.3 R_\odot) = 3.5 \times 10^5$ erg/cm^2/s if there were no damping. However, we have measured only 1.1×10^5 erg/cm^2/s. The difference in energy has been lost from the waves. Figure 4 illustrates this by showing our inferred v_{nt} scaled by the area expansion $A(R)/A(R_\odot)$. Thus, in Figure 4 a horizontal line indicates no damping.

The amount of energy carried by the waves appears to be sufficient to heat the coronal hole and drive the fast solar wind. Estimates for the amount of energy required typically calculate the energy injected at $1\ R_\odot$. On the basis of the area expansion factors (see also Fig. 4), we infer that the amount of energy in the waves at $1\ R_\odot$ in our measurements is $F \lesssim 1 \times 10^6$ erg/cm^2/s. The energy flux needed to heat a polar coronal hole was estimated in the 1970s to be $\approx 8 \times 10^5$ erg/cm^2/s.[46] However, because the solar minimum during this measurement was unusually quiet only 6×10^5 erg/cm^2/s would be required.[47] The upper bound we find for the energy flux of 1×10^6 erg/cm^2/s shows that there could be enough energy in the waves.

Wave damping

Because Alfvén waves are ubiquitous in the corona, they have been an attractive candidate for coronal heating even before observational evidence for wave damping was found. Consequently, a number of theories have been proposed for how the damping rate could be increased. These mechanisms rely on plasma inhomogeneities. Some examples include phase mixing due to gradients across the magnetic field and wave reflection and turbulence caused by gradients along the magnetic field.

One reason for large cross-field gradients is that the solar magnetic field is organized into magnetic flux tubes. These structures originate in the photosphere, where the magnetic field is concentrated into the boundaries of convection cells. Magnetic fields in these boundaries start as thin tubes of magnetic flux and expand at larger heights to fill the entire corona.[48] There can be large gradients in properties such as density and temperature at the boundaries between magnetic flux tubes.

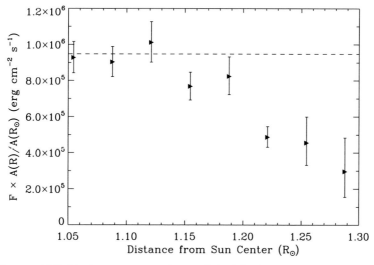

Figure 4. Energy flux F multiplied by the area expansion factor $A/A(R_\odot)$ for the Fe IX line. If the waves were undamped, then the points would fall on a straight line on this plot. The dashed line is drawn at the average of the three lowest points.

Phase mixing at these boundaries is one process that leads to increased damping.[49,50] Phase mixing can occur when there is a gradient in V_A across the magnetic field. The Alfvén speed is the phase speed of the waves. Thus, waves on neighboring field lines may be initially in phase with one another, but the different phase speeds cause them to become out of phase as they travel away from the Sun. Essentially, this causes friction between the waves. More precisely, phase mixing creates small length scales in the wave perturbations, which increases the rate of dissipative processes, such as resistive or viscous heating.

Along the field, the magnetic field strength and density, and hence the Alfvén speed, vary smoothly with height. If V_A changes slowly, that is, on length scales larger than a wavelength, then the waves can adjust to the changing properties of the plasma. However, if V_A changes on a scale shorter than the wavelength, then the waves can be reflected.[51] Alfvén waves with periods of order 300 s or longer are likely to be reflected in coronal holes.[51] The waves are predicted to have periods of 100–1,000 s, based on measurements of the photospheric convection that is believed to excite the waves.[52] Thus, some of these waves are likely to be reflected in the corona.

The interaction between the outward-propagating waves and inward-propagating reflected waves can cause the plasma to become turbulent. Turbulence causes the energy contained in the low frequency waves to cascade into higher frequency waves. These high-frequency waves damp rapidly through wave-particle interactions. In this way efficient heating is accomplished through the high-frequency waves with the low frequency waves acting as a reservoir of wave energy.[53] Observational evidence for this scenario in quiet Sun regions has recently been found through measurements showing that there is more power in the outward-propagating waves, which is consistent with turbulent dissipation.[54]

Ion cyclotron resonance heating is one example of how energy can be efficiently transferred between the high-frequency waves and the particles.[55,56] Ions in a magnetic field orbit the field lines at a frequency known as the ion cyclotron frequency $\Omega = qB/M$, where q is the ion charge, and M is the ion mass. Ion cyclotron resonance heating occurs when the frequency of the wave is close to the ion cyclotron frequency. In the corona, we expect that this will lead to more heating for ions with low charge to mass ratio q/M. This is because the turbulent wave spectrum follows a power-law as a function of frequency with more power in the low frequency waves.[57] Thus, it is predicted that T_i will be greater for low q/M ions.

We can estimate the ion temperature through v_{eff}. But because we do not have independent measurements of T_i and v_{nt}, we can only find upper and lower bounds for T_i.[58,59] To do this we assumed that all the ions have the same v_{nt}. This is because v_{nt} represents bulk motions of the whole fluid induced by the low frequency waves. To find an upper bound on T_i we set $v_{nt} = 0$ in Eq. (2), and to find a lower bound we use the condition $T_i \geq T_e$.[45]

Figure 5 shows the upper and lower bounds for T_i as a function of q/M in our data. The ion temperature depends on q/M with lower q/M ions having a higher temperature. This is evidence that there is ion cyclotron resonance heating and turbulence in the coronal hole. However, this does not necessarily prove that resonant heating is important compared to other heating processes. This is because the solar abundance of elements other than hydrogen and helium is very small, so it takes only a small amount of energy to heat these ions.

Conclusions and open questions

We now have unambiguous evidence that Alfvén waves are damped at low enough heights in the solar corona to be an important coronal heating mechanism. It remains to explain the physical reason why the waves are damped. A number of theories have been developed showing how the Alfvén wave damping rate might be increased due to the inhomogeneity of the corona. These ideas were illustrated above with phase mixing and wave reflection, but there are also other possibilities, such as through conversion of Alfvén waves into other types of waves that damp more easily[60] or through the absorption of surface waves at the boundaries of flux tubes.[61] All of these are examples of ways to generate small length scales in the wave perturbations that enhance the dissipation of wave energy into heat. How the structuring of the corona modifies wave properties is a subject of ongoing theoretical investigation.[62,63] Our work provides a motivation for more detailed theoretical calculations and a basis on which to compare them.

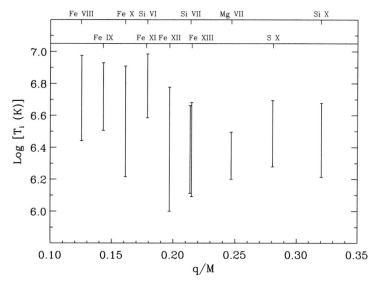

Figure 5. Upper and lower bounds for T_i as a function of q/M at a height of $\approx 1.06\ R_\odot$. The higher temperature of low q/M ions suggests ion cyclotron resonance heating of these ions.

Our estimated upper bound for the energy carried by these waves is sufficient for heating the observed coronal hole and accelerating the fast solar wind. Since we have found only an upper bound it is an open question whether reconnection is also important for coronal heating.

For this work we focussed on coronal holes, but other structures may be heated differently. Assuming that the energy flux of the waves is similar outside of coronal holes, our inferred flux of 10^6 erg/cm^2/s would also be sufficient to heat the quiet corona outside coronal holes, which require an energy input of about 3×10^5 erg/cm^2/s.[46] Active regions are another structure in the corona. They are areas of strong magnetic field near sunspots and are the source of solar flares and coronal mass ejections. These regions require an energy flux of 10^7 erg/cm^2/s, and so they could not be driven solely by waves, unless there is much more wave energy present in them than in the coronal hole.[46] Another consideration is that reconnection is more likely to be important for the quiet corona and active regions. Such regions consist of relatively short loops of magnetic field. The time it takes for a perturbation at a footpoint to propagate across the loop can be short compared to the timescale of the driving convection. This makes it possible to tangle the field lines, creating stressed magnetic fields that can reconnect.[64]

In summary, one of the major problems for wave-driven models of coronal heating has been resolved, namely to show that the waves dissipate energy at low heights in the corona. It is therefore likely that waves are an important factor in coronal heating, particularly for coronal holes and the fast solar wind. However, many open questions remain concerning the physical process behind wave dissipation, the role of reconnection, and the heating of other structures on the Sun.

Acknowledgment

The author would like to thank his post doc mentor Daniel W. Savin and collaborator Enrico Landi. This work was supported by a grant from the National Science Foundation Solar Heliospheric and INterplanetary Environment (SHINE) program.

Conflicts of interest

The author declares no conflicts of interest.

References

1. Foukal, P.V. 2004. *Solar Astrophysics.* Wiley-VCH. Weinheim.
2. Grotrian, W. 1939. Zur frage der deutung der linien im spektrum der sonnenkorona. *Naturwissenschaften* **27**: 214.
3. Edlén, B. 1943. Die deutung der emissionslinien im spektrum der sonnenkorona. *Z. Astrophys.* **22**: 30.
4. Chapman, S. 1957. Notes on the solar corona and terrestrial ionosphere. *Smithsonian Contrib. Astrophys.* **2**: 1.
5. Parker, E.N. 1958. Dynamics of the interplanetary gas and magnetic fields. *Astrophys. J.* **128**: 664.

6. Schwenn, R. 2006. Space weather: the solar perspective. *Living Rev. Solar Phys.* **3**: 2.

7. Pulkkinen, T. 2007. Space weather: Terrestrial perspective. *Living Rev. Solar Phys.* **4**: 1.

8. Zweibel, E.G. & M. Yamada. 2009. Magnetic reconnection in astrophysical and laboratory plasmas. *Annu. Rev. Astro. Astrophys.* **47**: 291.

9. Longcope, D.W. & L. Tarr. 2012. The role of fast magnetosonic waves in the release and converseion via reconnection of energy stored by a current sheet. *Astrophys. J.* **756**: 192.

10. McLaughlin, J.A., G. Verth, V. Fedun & R. Erdélyi. 2012. Generation of quasi-periodic waves and flows in the solar atmosphere by oscillatory reconnection. *Astrophys. J.* **749**: 30.

11. Klimchuk, J.A. 2006. On solving the coronal heating problem. *Solar Phys.* **234**: 41.

12. Cranmer, S.R. & A.A. Van Ballegooijen. 2010. Can the solar wind be driven by magnetic reconnection in the sun's magnetic carpet? *Astrophys. J.* **720**: 824.

13. Golub, L. & J.M. Pasachoff. 2010. *The Solar Corona.* Cambridge University Press. Cambridge.

14. De Pontieu, B. *et al.* 2007. Chromospheric Alfvénic waves strong enough to power the solar wind. *Science* **318**: 1574.

15. Tomczyk, S., S.W. McIntosh, S.L. Keil, *et al.* 2007. Alfvén waves in the solar corona. *Science* **317**: 5842.

16. McIntosh, S.W., B. De Pontieu, M. Carlsson, *et al.* 2011. Alfvénic waves with sufficient energy to power the quiet solar corona and fast solar wind. *Nature* **475**: 477.

17. Belcher, J.W. & L. Davis, Jr. 1971. Large-amplitude Alfvén waves in the interplanetary medium, 2. *J. Geophys. Res.* **76**: 3534.

18. Parker, E.N. 1991. Heating solar coronal holes. *Astrophys. J.* **372**: 719.

19. Cranmer, S.R. 2002. Coronal holes and the high-speed solar wind. *Space Sci. Rev.* **101**: 229.

20. Roberts, D.A. 2010. Demonstrations that the solar wind is not accelerated by waves or turbulence. *Astrophys. J.* **711**: 1044.

21. Kosugi, T. *et al.* 2007. The Hinode (Solar-B) mission: an overview. *Solar Phys.* **243**: 3.

22. Culhane, J.L. *et al.* 2007. The EUV imaging spectrometer for Hinode. *Solar Phys.* **243**: 19.

23. Hahn, M., E. Landi & D.W. Savin. 2012. Evidence of wave damping at low heights in a polar coronal hole. *Astrophys. J.* **753**: 36.

24. Young, P.R. *et al.* 2007. EUV emission lines and diagnostics observed with Hinode/EIS. *Publ. Astron. Soc. Japan* **59**: 857.

25. Brown, C.M., U. Feldman, J.F. Seely, *et al.* 2008. Wavelengths and intensities of spectral lines in the 171-211 and 245-291 å ranges from five solar regions recorded by the extreme-ultraviolet imaging spectrometer (EIS) on Hinode. *Astrophys. J. Suppl. Ser.* **176**: 511.

26. Fisk, L.A. & N.A. Schwadron. 2001. The behavior of the open magnetic field of the sun. *Astrophys. J.* **560**: 425.

27. Krieger, A.S., A.F. Timothy & E.C. Roelof. 1973. A coronal hole and its identification as the source of a high velocity solar wind stream. *Solar Phys.* **29**: 505.

28. Zirker, J.B. 1977. Coronal holes and high-speed wind streams. *Rev. Geophys. Space Phys.* **15**: 257.

29. Delaboudiniére, J.-P. *et al.* 1995. Eit: extreme-ultraviolet imaging telescope for the soho mission. *Solar Phys.* **162**: 291.

30. Domingo, V., B. Fleck & A.I. Poland. 1995. The SOHO mission: an overview. *Solar Phys.* **162**: 1.

31. Bryans, P., E. Landi & D.W. Savin. 2009. A new approach to analyzing solar coronal spectra and updated collisional ionization equilibrium calculations. II. updated ionization rate coefficients. *Astrophys. J.* **691**: 1540.

32. Beckers, J.M. & R.C. Canfield. 1975. Motions in the solar atmosphere. Tech. Rep. AFCRL-TR-0592, Air Force Cambridge Research Lab.

33. Deubner, F. 1976. "Dynamics of the solar atmosphere. Review of current observational investigations in the visible." In *Energy Balance and Hydrodynamics of the Solar Chromosphere and Corona: Proceedings of the International Astronomical Union, Colloquium no. 36.* R. Bonnet & P. Delache, 45. Clermont-Ferrand: G. de Bussac.

34. White, O.R. 1976. "High resolution observations of solar velocity fields from spacecrafts and rockets, using spectroscopic methods." In *Energy Balance and Hydrodynamics of the Solar Chromosphere and Corona: Proceedings of the International Astronomical Union, Colloquium no. 36.* R. Bonnet & P. Delache. Clermont-Ferrand: G. de Bussac.

35. Withbroe, G. 1976. "Mass and energy flow in the solar atmosphere—implications of skylab observations." In *Energy Balance and Hydrodynamics of the Solar Chromosphere and Corona: Proceedings of the International Astronomical Union, Colloquium no. 36.* R. Bonnet & P. Delache, 263. Clermont-Ferrand: G. de Bussac.

36. Doyle, J.G., D. Banerjee & M.E. Perez. 1998. Coronal line-width variations. *Solar Phys.* **181**: 91.

37. Banerjee, D., L. Teriaca, J.G. Doyle & K. Wilhelm. 1998. Broadening of Si viii lines observed in the solar polar coronal holes. *Astron. Astrophys.* **339**: 208.

38. Moran, T.G. 2001. Interpretation of coronal off-limb spectral line width measurements. *Astron. Astrophys.* **374**: L9.

39. Esser, R. *et al.* 1999. Plasma properties in coronal holes derived from measurements of minor ion spectral lines and polarized white light intensity. *Astrophys. J.* **510**: 63.

40. Moran, T.G. 2003. Test for Alfvén wave signatures in a solar coronal hole. *Astrophys. J.* **598**: 657.

41. Dolla, L. & J. Solomon. 2008. Solar off-limb line widths: Alfvén waves, ion-cyclotron waves, and preferential heating. *Astron. Astrophys.* **483**: 271.

42. Bemporad, A. & L. Abbo. 2012. Spectroscopic signature of Alfvén waves damping in a polar coronal hole up to 0.4 solar radii. *Astrophys. J.* **751**: 110.

43. Wang, Y.-M., E. Robbrecht & N.R. Sheeley, Jr. 2009. On the weakening of the polar magnetic fields during solar cycle 23. *Astrophys. J.* **707**: 1372.

44. Cranmer, S.R. *et al.* 1999. An empirical model of a polar coronal hole at solar minimum. *Astrophys. J.* **511**: 481.

45. Hahn, M., P. Bryans, E. Landi, *et al.* 2010. Properties of a polar coronal hole during the solar minimum in 2007. *Astrophys. J.* **725**: 774.

46. Withbroe, G.L. & R.W. Noyes. 1977. Mass ane energy flow in the solar chromosphere and corona. *Annu. Rev. Astro. Astrophys.* **15**: 363.

47. McComas, D.J., R.W. Ebert, B.E. Goldstein, *et al.* 2008. Weaker solar wind from polar coronal holes and the whole sun. *Geophys. Res. Lett.* **35**: 18103.

48. Spruit, H.C. & B. Roberts. 1983. Magnetic flux tubes on the sun. *Nature* **304**: 401.

49. Heyvaerts, J. & E.R. Priest. 1983. Coronal heating by phase-mixed shear Alfvén waves. *Astron. Astrophys.* **117**: 220.

50. Hood, A.W., J. Ireland & E.R. Priest. 1997. Heating of coronal holes by phase mixing. *Astron. Astrophys.* **318**: 957.

51. Moore, R.L., Z.E. Musielak, S.T. Suess & C.-H. An. 1991. Alvén wave trapping, network microflaring, and heating in solar coronal holes. *Astrophys. J.* **378**: 347.

52. Cranmer, S.R. & A.A. Van Ballegooijen. 2005. On the generation, propagation, and reflection of Alfvén waves from the solar photosphere to the distant heliosphere. *Astrophys. J. Suppl. Ser.* **156**: 265.

53. Matthaeus, W.H., G.P. Zank, S. Oughton, *et al.* 1999. Coronal heating by magnetohydrodynamic turbulence driven by reflected low-frequency waves. *Astrophys. J.* **523**: L93.

54. Tomczyk, S. & S.W. McIntosh. 2009. Time-distance seismology of the solar corona with comp. *Astrophys. J.* **697**: 1384.

55. Cranmer, S.R., G.B. Field & J.L. Kohl. 1999. Spectroscopic constraints on models of ion cyclotron resonance heating in the polar solar corona and high-speed solar wind. *Astrophys. J.* **518**: 937.

56. Hollweg, J.V. & P.A. Isenberg. 2002. Generation of the fast solar wind: a review with emphasis on the resonant cyclotron interaction. *J. Geophys. Res.* **107**: 1147.

57. Leamon, R.J., C.W. Smith, N.F. Ness, *et al.* 1998. Observational constraints on the dynamics of the interplanetary magnetic field dissipation range. *J. Geophys. Res.* **103**: 4775.

58. Tu, C.-Y., E. Marsch, K. Wilhelm & W. Curdt. 1998. Ion temperatures in a solar polar coronal hole observed by SUMER on SOHO. *Astrophys. J.* **503**: 475.

59. Landi, E., & S.R. Cranmer. 2009. Ion temperatures in the low solar corona: Polar coronal holes at solar minimum. *Astrophys. J.* **691**: 794.

60. Uchida, Y. & O. Kaburaki. 1974. Excess heating of corona and chromosphere above magnetic regions by non-linear Alfvén waves. *Solar Phys.* **35**: 451.

61. Ionson, J.A. 1978. Absorption of Alfvénic surface waves and the heating of solar coronal loops. *Astrophys. J.* **226**: 650.

62. Goossens, M., J. Terradas, J. Andries, *et al.* 2009. On the nature of kink mhd waves in magnetic flux tubes. *Astron. Astrophys.* **503**: 213.

63. Goossens, M., J. Andries, R. Soler, *et al.* 2012. Surface Alfvén waves in solar flux tubes. *Astrophys. J.* **753**: 111.

64. Parker, E.N., 1983. Magnetic neutral sheets in evolving fields. II. formation of the solar corona. *Astrophys. J.* **264**: 635.